IFAC WORKSHOP ON DISTRIBUTED COMPUTER CONTROL SYSTEMS 1997

Sponsored by :
IFAC – International Federation of Automatic Control
Technical Committee on Distributed Computer Control Systems(CCD)

Organized by :
Engineering Research Center for Advanced Control and Instrumentation(ERC-ACI), Seoul National University

On behalf of
Institute of Control, Automation and Systems Engineers(ICASE), Korea

International Program Committee

Ian MacLeod, Chairman(South Africa)
Wook H. Kwon, Vice-chairman(Korea)
Mike G. Rodd(UK)
Kirithi Ramamritham(USA)
Hermann Kopetz(Austria)
Heonshik Shin(Korea)
S.Joe Qin (USA)
Kang G.Shin(USA)
Jin S. Lee(Korea)
Juan A. de la Puente(Spain)
L.Boullart(Belgium)
Baosheng Hu (PR China)
M.A.Inamoto(Japan)
Kader Sahraoui (France)
Joo-Kang Lee(Korea)
Arcot Sowmya(Australia)
S.Narita(Japan)
Frnacesco Tisato(Italy)

National Organizing Committee

Wook H. Kwon, Chariman,
Ic-Soo Park
Seung-Ho Hong
Hon-Seong Park
Seong-Soo Hong
Jaehyun Park
Seong-Hwan Kim
You-Keun Park
Sang-Keun Shin

PREFACE

It is increasingly accepted that future dependable, real-time digital computer control systems will have distributed architectures. Advantages of distributed computer control systems include the possibility of composing large systems out of pre-tested components with minimal integration effort, their well-defined fault containment properties and their capacity to make effective use of mass-produced silicon chips.

The IFAC Workshop series on Distributed Computer Control Systems (DCCS) focuses on design requirements and fundamental principles encountered in such systems and highlights and traces the growth of key concepts at their various stages of development. Theoretical and application-oriented viewpoints receive equal emphasis. These Workshops also provide an excellent forum for the exchange of information on recent technological advances and practices in the distributed computer control field.

The 1997 DCCS Workshop was notable for the attention given to practical implementations of ideas that have been under discussion for decades. Probably the most significant was experience with achieving constructive, component-based design of complex control systems. A prerequisite for this approach to system building is that validated properties of components must not be affected by system integration. Time-triggered, rather than event-triggered approaches to temporal data and its management have shown significant advantages in this respect. However, changes in operating conditions or the mode of operation of the physical system being controlled can require a more flexible approach. Some favour using event-triggered mechanisms for this, but alternative approaches were also suggested. Protocols for achieving validity and consistency in the face of failures such as a "babbling idiot" node computer still pose significant challenges.

True real-time communication systems also received much attention and practical implementations were discussed. It was generally agreed that in order to produce such systems, rigorous specification, design and verification processes are essential. However, to achieve this, a mixture of tools is currently necessary. Simpler, object-oriented tools will be required in the future.

Fieldbus systems continue to be of great interest and the results of experiments and practical implementations using existing, commercial communication systems were reported. Unfortunately, it appears as if none of the presently-available systems meets all the requirements. However, several interesting proposals were discussed and the recent appearance of cost-effective, powerful silicon chips will have a big influence on future intelligent sensors, communication systems and their economics.

Finally, applications reported included control systems for steel production, electricity generation, integrated control of commercial ships, robot navigation and home automation.

The IFAC DCCS Workshops continue a tradition started in the 1970s. They have maintained a consistently high technical standard. This 14th Workshop was no exception - the policy of concentrating on a specific topic, inviting a number of key authors and of accepting only a limited number of papers has paid dividends. In the Workshop situation, this leads to lively and fruitful discussion that is the highlight of the entire exercise. The venue, Seoul National University, provided an ideal environment for such debate. Seoul, the capital city of Korea for the past 600 years, now the modern centre of this newly-industrialized country, provided an ideal base, facilitating the attendance of approximately 40 Workshop participants from Asian countries plus 15 participants from 10 other countries. It is the Editors' hope that readers of the papers contained in these Proceedings will be able to share and benefit from the stimulating exchange of ideas.

Prof. I.M. MacLeod and Prof. W.H. Kwon,
Editors.

CONTENTS

REAL-TIME SYSTEMS

MODELING FOR DISTRIBUTED COMPUTER CONTROL SYSTEMS

APPLICATIONS IN DISTRIBUTED COMPUTER CONTROL SYSTEMS I

FIELDBUS NETWORK FOR REAL-TIME CONTROL

REAL-TIME NETWORK FOR DISTRIBUTED CONTROL

APPLICATIONS IN DISTRIBUTED COMPUTER CONTROL SYSTEMS II

KEYNOTE PAPER

ARCHITECTURE FOR DISTRIBUTED CONTROL SYSTEMS

THE DESIGN AND ANALYSIS OF LOW-COST REAL-TIME FIELDBUS SYSTEMS

M G Rodd* , K Dimyati ** and L Motus***

*University of Wales Swansea, Singleton Park, Swansea SA2 8PP, UK.
Email: M.G.Rodd@swansea.ac.uk
**Universiti Malaya, 50603 Kuala Lumpur, Malaysia.
***Tallinn Technical University, Estonia

Abstract: Extensively discussed for decades, it is only relatively recently that a demand for true real-time industrial communication systems has become evident. However, when the requirements of such systems are understood, few of the offered solutions are appropriate. This paper suggests that, in order to produce such systems, rigorous specification, design and verification processes are necessary, and that these must all be capable of handling logical and temporal analysis and design. To achieve this, it is suggested that a mixture of tools is (currently) necessary, and that these tools should include object-oriented design techniques with temporal extensions. The points are illustrated by reference to the recent development of a real-time fieldbus system, built around the CAN protocol, which supports true real-time communication over an existing, commercial system.

Keywords: Fieldbus, communication systems, communication protocols, real-time communications, clock synchronization.

1. INTRODUCTION

The benefits of having available universally-accepted communication systems for use in factory and process automation are well understood and, on the surface, are well accepted by suppliers and users alike. The fact that in practice little has been achieved in producing such generic products, though, must give rise to some concern as to the real enthusiasm of both parties. Where success has been achieved, such as in the automobile and aerospace industries, there are some very clear driving forces – which, maybe, are not present in the more-general areas of automation.

The problems to be faced are by no means trivial, however, as there are several, often conflicting, forces at work. The first is the relatively small size and cost structure of the more generic automation business. The total control and instrumentation budget for a typical large industrial project might only amount to a few percentage of the total cost. Communication systems thus rank very low in the available budget. The market for commercial data-communication systems is vast. The natural temptation, therefore, is to adopt systems developed for the commercial sector, in the industrial sector. However, the fundamental needs of the two areas of application are fundamentally different – as will be pointed out later.

The next issue is that, given this relatively small market, companies who presently have successful products are not likely to really back the development of new products that will impact the sales of their own well-developed and well-respected existing offerings. Of course, in public, they must show their willingness to see open standards adopted, but in private they have businesses to run, and shareholders to satisfy!

Finally, from a more technical point of view, given the rapid march of electronics and communications techniques – offering, for example, the promise of literally limitless bandwidth at almost zero cost – do any of the current proposals offer anything more than mere interim solutions? Can any sensible technical director put a company's hard-earned profits into technologies which seem to offer capabilities that are already orders of magnitude less powerful than the simple connection to a desk-top computer?

1

Despite these negative comments, there is a force for change, and it is essential that in the move towards standards-based industrial communication systems, designers and users alike appreciate their fundamental, desired characteristics. This paper will not argue the commercial case or undertake yet another debate about the merits of the current offerings. (See, for example, (Reeve *et al.*, 1997).) Rather, the paper will address one specific aspect of the requirements for such future products. This is their real-time performance. The standpoint taken is that it is increasingly being recognised that such systems must operate in a true real-time environment. To achieve this end, it is essential that the fundamental characteristics of the communication facilities provide the requisite support. The paper discusses these attributes, and illustrates steps that can be taken to match them in practice. To illustrate the approach, the practical development of a system based on one of the more successful fieldbus systems, CAN, is described.

2. "REAL TIME" IN INDUSTRIAL COMMUNICATION SYSTEMS

The IFAC DCCS workshops have played a major role in developing, and transferring to industry, a deep understanding of the inherent nature of real-time systems. The recent text by one of the pioneers of this workshop series, Hermann Kopetz (1997), provides probably one of the very best starting points for an understanding of the inherent nature of such systems – from where the development of industrially relevant solutions can start.

The key is simple: a real-time system must guarantee its performance in both the logical and the temporal domains. A correct answer provided late is as useless as the wrong answer delivered at the right time. The details of the timing requirements are thus always determined by the application, and not by the computing system – or any of its component parts. To this end it is essential that any real-time computing system must provide predictable and explicit processing – in terms of both correct data handling and execution timing. This processing performance must be end-to-end – from the lowest-level, isolated sensor, right through all layers of the hardware and software (including all software components), and back to an isolated actuator.

The consequence is that in developing any real-time system, all the processing aspects of all the system components must be designed and proved to be both logically and temporally deterministic. Here, "deterministic" is taken to mean that the performance of the object of interest is fully defined and understood. This naturally raises many issues, in that it will often not be possible to have complete knowledge of the environment being supported by the computing system. The point is that this lack of knowledge is no reason to make the supporting system non-deterministic!

To ensure this determinism, all the components of the designed system must have fully defined performance – and this must include the communications facilities.

It is also appropriate here to point out that in terms of performance of the communication systems, attention must be given to end-to-end performance. Thus, whilst MAP, for example, uses a 10 Megabits/second physical communications channel, and token-passing for media access, once all the layers of the associated protocols have been executed, the resulting performance is often in the low kilobits/second range, accompanied by totally unpredictable performance! It is also important to recognise that, contrary to the situation in most commercial data-processing systems, simply guaranteeing the maximum throughput bounds is not sufficient in real-time systems. An early result can often be as disastrous as a late one – as sampling theory proves! The consequence is that both upper and lower performance bounds must be guaranteed.

These bounds must, as mentioned previously, be total end-to-end performance bounds. Thus, all aspects of the system between the end points, including all aspects of software processing, communications protocol handling and the physical transfer of the bits, must be considered. This is not a trivial problem, given that it is still not possible to measure the processing time of even relatively simple processors! (See (Olyeni, 1996) for example.) Further complications occur when it is appreciated that most, if not all, current "popular" computer architectures are incapable of providing guaranteed execution times.

Another aspect of real-time systems is the inherent need for a globally agreed time reference. This, again, has been well motivated by Kopetz and others. Whilst solutions such as those that make use of GPS have been proposed, there are powerful arguments for providing such agreed clock facilities across the communication channels. One simple reason for this is the basic problem of ensuring fault tolerance in the distributed clock system.

Clock synchronisation can be defined as a mechanism to co-ordinate, or unify, the notion of time across a distributed real-time system, so that the representation of time, anywhere in the system, is consistent. This is vital when organising and distributing shared resources – such as materials, tools, and information. In an automated factory, for example, one of the most important issues in the co-ordination of flexible manufacturing systems (FMS) is synchronisation amongst the processes, which can largely be based on a common, agreed timing reference. Another example is found in real-time distributed processing systems, such as those used in factories, aircraft, space vehicles, life-support systems and military applications. Failure to synchronise the clocks to a tolerated precision, in the case of a life-support system for example, could cause life to be in danger. When clocks are not synchronised correctly, it is impossible to impose strict timing relations, except by the extensive use of local sensors and interlocking.

3. APPROACHING THE DESIGN OF A REAL-TIME COMMUNICATION SYSTEM.

Once one recognises the complexity of developing a real-time communication system, it becomes evident that each and every aspect of the performance must be examined. As is well understood in control-systems design, it is vital that proposed systems are analysed at all stages of the design. In the case of sheer overall data-handling performance, and for verifying protocol performance, various discrete modelling tools such as BONeS (BONeS, 1993) can be used to model certain aspects of a proposed communications system. Such modelling is clearly very important in analysing the logical performance of the proposed system. However, a detailed temporal performance analysis is not possible, particularly in that all causal relationships, and hence, implementation details, need to be fully described.

However, in developing supporting software, and then analysing total system performance, far more extensive analysis is required. The steady flow of new software development support tools, the majority of which claim to support the development of time-constrained software, has not yet provided any widely approved methodology and/or industrially accepted tools. It is suggested that the problem cannot be solved by trying to adopt a smooth evolution of existing computational models and methodologies, and that new approaches are necessary.

In order to address the problems, it is necessary to understand the fundamental nature of real-time systems. Motus and Rodd (1994) have studied several operational real-time computer control systems and have pointed out the following specific features, present in all the systems:

- Forced parallelism, imposed and controlled by the environment, in addition to common parallelism, introduced by the designers and controlled by the computer programs themselves.
- Synchronous, semisynchronous and asynchronous execution modes of computational processes.
- The necessity for time-selective interprocess communication.

All these considerations are fundamental when designing end-to-end communication systems. The implications of forced parallelism and time-selective interprocess communication – although obviously widely known, since they are present in the majority of implemented systems – have not gained sufficient attention from theoreticians. As explained in (Motus, 1995), in order to study forced parallelism, it is necessary to change the conventional pattern of thinking about real-time software. This change leads to computational models where real-time software is considered to be a collection of autonomous, loosely coupled, interacting, repeatedly activated and terminating programs, instead of being one single non-terminating program.

The suggested paradigm assists in the time-labelling of all the data items used in a system, and in the explicit definition of activation instants for each program (if necessary). The software designer of a time-critical system can then concentrate on determining the necessary, time-related properties of the software, and how these properties may be verified. Three groups of timing properties are emphasised in (Motus and Rodd, 1994) – performance-bound properties, timewise correctness of events and data, and time correctness of interprocess communication. In practice, the performance-bound properties are normally covered, but the other two groups of properties are only studied occasionally, and their handling is catered for by only a very few CASE tools.

Current practice in the handling of the timing properties (say, using Petri-net-based tools) allows for the detection of the more obvious errors (e.g., deadline violation of a response time). Such errors can also be detected by simulation. However, more subtle, maybe malicious, timing errors can remain undetected – for example, a too-long non-transport delay in the interprocess communication between two asynchronous processes (Motus and Rodd, 1994). Since such a delay behaves with a saw-tooth time function, the resulting excessively long delays will occur at seemingly random time instants and, hence, will be practically impossible to detect by testing. The only way to find such subtle timing errors is by an analytical study based on a formal description of the software.

Another common misconception is that timing errors are regarded as run-time errors. The truth is that, like most software errors, the bulk of timing errors stem from incorrect, or un-stated, user requirement specifications. In many cases such errors arise from implicit assumptions made in the underlying theory or understanding of a system, or from a conflict between such assumptions. For example, a sampled data control system that has been proved to be stable, may become unstable if the regularity of the time between the sampling of data is violated (for example, see (Törngren, 1995)). There can be no doubt that a time irregularity in the measurements is a timing error when regularity is required and assumed during controller design. Equally, the sources of timing errors may often arise from inconsistent information, given by different experts, and used in putting together the requirements.

From a careful study of experiences in the implementation of various real-time systems one can learn the following lessons:

- Timing errors are not necessarily only run-time errors, and cannot be detected and eliminated only by applying an appropriate scheduling discipline.
- Many timing errors can be detected only by analytical study of the system's properties, before the implementation stage is even reached.
- To achieve time-correct behaviour of a system, all three groups of timing properties should be proved.
- The search for timing errors (verification of time correctness) should start from the user requirements, and should be an on-going activity throughout the development cycle of the software.

As was mentioned earlier, there is currently no generally recognised integrated methodology, theory and/or CASE tool that can handle all timing considerations. To meet this need, this paper discusses a new, unified approach, based on the use of a new CASE tool called LIMITS. This tool is based on a technique that is mathematically complete, and hence supports full formal analysis.

3.1. The background to LIMITS

The new CASE tool, LIMITS, has been developed under an EU-funded trans-European research contract, as a unifying link between the object-oriented modelling and design methodology (Rumbaugh, et al., 1991) and the Q-model. The latter is a formal method that adopts the ideology of abstract communicating processes, and focuses specifically on the full temporal verification of designs throughout their lifecycles.

Successful attempts to build unifying links between various research domains have been published before – for instance, HRT-HOOD links scheduling theory with hierarchical object-oriented design (Burns and Wellings, 1995). Unlike the HRT-HOOD approach – which provides a new, independent CASE tool – LIMITS does not physically modify the associated OMT tools. Instead, LIMITS utilises designs produced by the OMT tools, transforms them according to its own Q-model needs, verifies their temporal correctness, allows the users to modify the designs (if necessary), and finally feeds the verified designs back to the OMT tool. At this stage, the interfaces between the OMT tool and LIMITS are, to a large extent, manual. This means, for example, that there is a need for the manual introduction of additional temporal attributes, constraints and requirements when transforming an OMT object model into its equivalent Q-model, and also when returning a verified Q-model to the OMT toolset.

3.2 The Q-model

The Q-model is based on a modified definition, first introduced by Quirk and Gilbert (1977), of a mathematical mapping that can be used to describe a process in a computational model. This modification enables the full time-labelling of all data items used in a system, and also the handling of each process as an autonomous entity inside a system of co-operating processes. The Q-model consists of processes, described as "mappings", which interact with each other via channels, also described as mappings.

The interaction of processes is time-selective. The time-selectivity is guaranteed by a channel function which defines the producer's process state values that are accessible via the channel, as a function of the consumer process activation instants. This results in a request for a message from the channel. In other words, the consumer process can subscribe to data on a time-selective basis.

A detailed description of the Q-model is given in (Motus and Rodd, 1994). The major advantage of the approach is its ability to capture time requirements and constraints at the requirements specification phase, and to analyse the specification mathematically for timing errors. However, the same formalism can be used to analyse the overall physical design and implemented software. The Q-model provides for three philosophical concepts of time (logical, absolute and relative), and this allows the approach to cover a complete range of all possible timing properties.

Another important aspect of the Q-model approach is that it handles periodic and aperiodic (sporadic) processes in the same way. This is achieved by introducing the notion of a random period. Interval estimates are used, for pragmatic reasons, instead of random variables and probability densities. This enables, for example, the consideration of the minimum and maximum frequencies of activations of sporadic processes.

Important in the development of distributed and/or safety-critical applications is that each process may be considered in the Q-model, in principle, as a fully autonomously executed entity. Each process may be activated by its own clock, or by the occurrence of a specified event – in the environment or in the computer system. Therefore, three synchronization modes are available in the Q-model – synchronous, semisynchronous and asynchronous. This requires the use of different types of channels to describe the interactions of processes executed in different modes. The temporal properties of interactions between processes executed in different modes can be analysed in the Q-model. For details see (Motus and Rodd, 1994).

3.3. Timed object modelling (TOM)

The Q-model can be, and indeed has been (see (Motus and Rodd, 1994)), used for implementing a stand-alone CASE tool. However, due to its formal nature, such a tool presents difficulties for its users when they move from the informal world of applications to the formal world in which the specification and design can be verified. Similar difficulties – often called the semantic gap – seem to hinder the wider acceptance of formal methods by the target users. As a result, a major decision was made to associate the Q-methodology with a standard set of CASE tools. An intense study of the techniques actually being used by the practising engineering profession showed that the OMT-based techniques had gained serious acceptance. Thus, it was decided to unite the Q-method with a standard implementation of OMT – and the result was LIMITS! This required some re-thinking of the approach, and an understanding of the tools required at each stage in the design.

A *system Q-model* is a complete model of the system under study (of its specification, design or implementation). It is a compiled set of class Q-models, object processes and even (perhaps) of separately described regular Q-model processes and channels, so that all the interface ports have been eliminated. In order to be able to analyse (and animate) the system Q-model, it should be composed of processes communicating via channels, and with all necessary timing characteristics, actual or estimated, known.

The *heredity tree* copies to a large extent the aggregation hierarchies from the object model. Any differences from the object model hierarchy appear because instances of classes are described explicitly, and some of the operations may be represented by their decompositions in order to better describe their behaviour.

A designer using the LIMITS approach to timed object modelling (see (Naks, 1996)) starts by creating an object model, as in OMT. The system is described in object classes which are then decomposed as necessary. After the system has been partitioned into object classes, one may start with the description of the inner structure and behaviour of the classes. The designer may also create functional and dynamic models, and may make use of features such as event traces, USE-CASES and scenarios in developing the system model.

The object model is then taken as the starting point for building the Q-model of the system. It has been found useful to start by describing the class Q-models and object processes. Then, with the help of the heredity tree, the system Q-model can be formed. The dynamic and functional models, plus any relevant additional information, are used as explanatory background information. Use-cases, event traces, and state diagrams help to capture knowledge about the execution sequences of processes, and the logical constraints under which these executions occur. The functional model also gives some insight into the data flows exchanged between processes. Quantitative time requirements and constraints usually require separate investigation and elicitation efforts, because they are not present in OMT models.

It is important to stress that all the initial time constraints and requirements are largely determined by the application, and serve as source information determining the requirements on the physical implementation of the system. The requirements for scheduling in the final implementation and execution of the physical design are formed later on in the development process. The knowledge of other non-functional requirements is also extremely useful because of their strong influence on the timing correctness of the system. The temporal analysis of the system's description can start at this stage; properties of single processes may be analysed as soon as the processes are described.

3.4 Simulation of system's behaviour

Formal analysis can only prove that the system's description is consistent and has certain required properties. The correspondence between the system's description and the actual real-world application cannot be verified formally. This can only be demonstrated by using informal simulations of the system's description, based on "what-if" scenarios. The results of these simulations must then be compared with the expected behaviour of the application in the real world. LIMITS combines formal and informal studies of the system, so that they support each other. When performing instance-level analysis, a prototype version of the system is automatically generated. Simulation can thus be carried out on the prototype to check hypotheses, stated as "what-if" scenarios, and supported by pre-specified, default, simulation modes. The simulation can be performed on the prototype based on the specified parameter values, or some of the timing parameters can be changed to study the influence of new values on the system's functioning. Designers may study the simulation results visually on time diagrams, or can use their own analysis programmes.

4. THE PRACTICAL DESIGN AND IMPLEMENTATION OF A REAL-TIME FIELDBUS

To illustrate some of the techniques and tools necessary in the development of a real real-time communication system, the rest of this paper will describe a system recently produced in Swansea. This system, a commercially driven development, has resulted in a unique real-time fieldbus, built around a presently available product. The resulting system, now fully tested, is ready for full integration.

4.1. CAN as an industrial fieldbus

The steady increase in factory automation systems has created a need for reliable, low-cost, real-time communication systems, which, amongst other things, resulted in the so-called "fieldbus" technology. It was in the late 1970s that this technology began to attract industrial attention, as it was perceived by industrialists as addressing real-time communication issues at the sensor-actuator level. As a result, there are many international exercises that are attempting to provide the "best-fit" solution to this low level, real-time application problem. Vendor-proprietary protocols such as HART, and regional standard protocols such as FIP, PROFIBUS and CAN, are amongst the contenders. The trick is to choose the "right" technology for the problem at hand. However, whether any of the competitors offer the right solution is still not clear – but, as mentioned previously, that debate will not be entered into here!

CAN was chosen in this present work mainly because of its wide acceptance, acceptable speed, high noise immunity, low cost per node, and relative degree of determinism. However, as pointed out above, for a fieldbus such as CAN to provide for truly real-time communications, it must be deterministic, and must also be able to cope with various time constraints imposed upon it by the attached processes. Thus, as one of the requirements, the fieldbus must incorporate a mechanism for maintaining a consistent global time. To support this, real-time clock synchronisation is required (Dimyati and Rodd, 1995).

CAN, the Controller Area Network protocol (ISO 11898, 1993), is a serial communication protocol, developed by Robert Bosch GmBH for use in time-critical applications in the automotive industry. It has been designed to operate in noisy environments, at speeds up to 1 Mbit/s (if the bus length is less than 40 metres). CAN uses a non-destructive bitwise arbitration to decide which node "owns" the bus at any instant of time, and this involves a message priority scheme that is based on the value of the message identifier transmitted with each message. This provides a (partial) solution for ensuring that urgent messages get through, and makes it possible (under severe restrictions) to guarantee real-time requirements. Following its successful implementation in the automotive industry, CAN has been found increasingly useful in the more-general industrial automation field. This migration has basically been driven by its inherent characteristics – largely its robustness, in conjunction with its excellent performance/price ratio (Lawrenz, 1995). The high sales, of more than 9 million units up to 1994, and the many promising forecasts, illustrate the importance of CAN as a future protocol for use as an industrial fieldbus (Lawrenz, 1995).

4.2. Establishing global time in CAN

CAN does not, at present, have any direct facilities to support real-time clock synchronisation amongst the interconnected nodes. It was therefore the aim of the work described in this paper to provide a global, system-wide time reference for CAN, as part of a larger exercise on advanced intelligent manufacturing systems.

To make the clock synchronisation cost-effective, a hybrid approach to clock synchronisation, based on a hardware-assisted software technique, was adopted. The system developed (see Fig. 1) consists, essentially, of:

- a dedicated, but very simple, microcontroller, which is used for processing the clock-synchronisation algorithm, and
- the use of the TICK chip – a specially designed clock-synchronisation hardware unit (TICKER, 1994) which includes facilities such as an adjustable clock, control logic and a local processor interface.

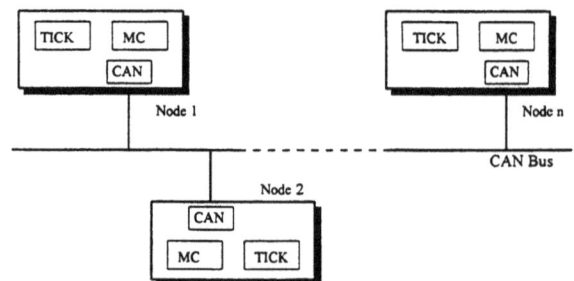

Fig. 1. Basic system architecture

The design set out to satisfy the desired features that had been defined at the requirements specification stage:

- The prototype system had to be integrated into a low-cost PC-based evaluation system.

- The resulting system had to provide the system with a simple networking capability based on a polling approach, without the use of any system-wide interrupts.
- The clock-synchronisation subsystem would not be permitted to have any significant impact on other system activities. In particular, it should not compete for system resources, such as the main CPU time.
- The network access associated with the clock-synchronisation process should not interfere with application-specific network activities.
- The achievable accuracy of the globally synchronised time should be optimal, and should be limited only by the minimum bit time of the CAN bus (1 μs).
- The system had to be provided with elementary embedded software debugging support, as it was to be used for development purposes.

In addition, the final product had to be, at least potentially, easy to integrate into a single chip.

In meeting the basic design requirements, the TICK chip (a patented development, now commercially available) was used (TICKER, 1994). The use of this chip reduced the design time for developing the time-synchronisation unit. The chip can maintain network synchronisation under software control, and makes the time reference available to the application processor, either through one of the four "snapshot" registers, or through elapsed time, with the use of a so-called "time bomb" (TICKER, 1994). Time, from the point of view of the TICK chip, is represented by a 56-bit unsigned integer measured with a granularity of microseconds. In order to set the time value in the TICK chip, there are 8 registers – one for the "flag byte" and seven for the clock values. The flag byte is used for administrative purposes by the software, and amongst other things, carries the "accuracy" value, so that it can be checked at any time to ensure that the system is well synchronised.

On the software side, a fault-tolerant midpoint algorithm (Welch and Lynch, 1988) was used. This algorithm provides a fast and efficient estimation of the approximate global time. With this carefully selected synchronisation algorithm (as a hybrid clock-synchronisation approach was adopted), a practical, working solution was produced. The prototype real-time CAN board is as shown in Fig. 2. The key aspect of the resulting architecture is the use of a separate processor to handle all the synchronisation processing. It is considered very important that there is no interference with the work of the local processors.

The resulting system was extensively tested in a laboratory environment, and was shown to be a highly effective solution. The current implementation uses a 4-layer PCB plug-in card, which provides any related application with simple, continuous access to the globally synchronised time.

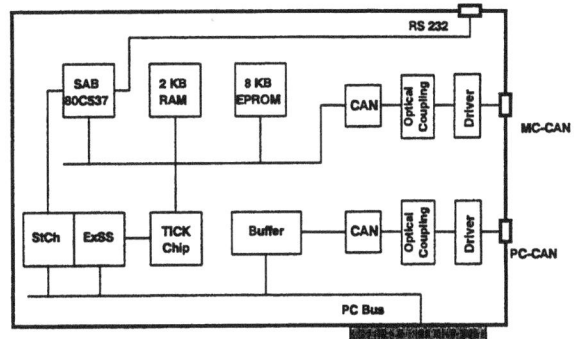

Fig. 2. The prototype real-time CAN board

It should be noted that, because the microprocessors involved do not normally support multiple-processor shared bus access, it was not possible to share the same CAN controller between the main CPU (Intel 80x86-compatible) and the on-board microcontroller (Intel 8051-compatible). Thus, to guarantee full transparency of the clock-synchronisation process, two CAN controllers were used. One was allocated exclusively to clock-synchronisation network traffic, and the other was interfaced to the PC-host for handling system messages.

A similar device-sharing problem was encountered with the specialised clock-synchronisation TICK chip. On the one hand, this has to be programmed from the local processor executing the clock-synchronisation algorithm. On the other hand, though, it has to allow the PC to access the synchronised time. Fortunately, the design of the TICK chip provides an additional multiplexed time bus output, which can be decoded and interfaced to the PC bus through a specially designed "External Snapshot" register (ExSS) (Dimyati, 1996). This is implemented via an EPLD (Erasable Programmable Logic Device) chip.

As can be seen from Fig. 2 above, all the devices in the board are interconnected via two buses. The first one, MC-CAN, is driven by the microcontroller, the SAB 80C537, and this is dedicated to the clock-synchronisation subsystem. The second bus, PC-CAN, is directly connected to the PC bus, and allows for CAN network access, and for the reading of the current time from the External Snapshot register. These two buses are connected via the registers of the Configuration/Status Channel, which makes it possible to perform basic communication between the PC and the clock-synchronisation subsystem. To protect the board from the effects of external voltage surges, and to avoid ground loops, both CAN controllers are optically de-coupled from the network.

7

5. PRACTICAL RESULTS

Following the initial block-diagram-level design of the proposed real-time CAN system, extensive logical modelling was first undertaken to prevent large amounts of design time being wasted. This work is discussed in (Dimyati, 1996). The results of this exercise were extremely valuable, in that they showed that it was feasible to achieve the required throughput, even in the presence of the required clock-synchronization messages. Also, aspects of the logical performance of the system could be fully understood. On the basis of this preliminary evidence, more detailed design and analysis could take place.

Although the full use of LIMITS was not possible at that stage (LIMITS and the real-time CAN projects being undertaken in parallel), sufficient time was available to analyse certain aspects of the new communications system. This work was also assisted by earlier studies on protocol modelling using the Q-model approach, implemented in an earlier version of LIMITS (called CONRAD) and reported in (Motus and Rodd, 1994). The result of all of this was that before the prototype hardware system was built, performance could be predicted, and appropriate hardware chosen. The final operation of the CAN-based system did, in fact, prove to support all the prior analysis (Dimyati, 1996).

Although very complex, the experimental system proved to be relatively straightforward to develop and test. The real-time CAN boards were plugged into the PC-host stations and connected via a CAN network. The hosts were used primarily for configuration purposes, network traffic monitoring and software development. The binary modules of the clock-synchronisation software were loaded into the EPROM and transferred into the EPROM sockets. The real-time CAN boards were also connected to the serial RS 232 ports of the PC which act as terminal emulators. The software modules were arranged so as to provide easily readable status and debugging information via these ports, allowing for the rapid inspection of results.

The real-time CAN boards were operated continuously in a laboratory environment, and the clock value captured for every re-synchronisation interval was recorded in a file so that results could be evaluated. In order to spread the adjustment of the clock correction over a relatively long period, the lowest slew rate of about 7 ms/s (based on the TICK specifications (TICKER, 1994)) was chosen for use during the stable state. A higher rate can also be selected, but the slewing will then be operated over a relatively short period, thus leaving a longer period during which no slewing occurs – i.e. the adjustment will take place over a very small part of the re-synchronisation interval, and no adjustment will be made in the remaining part of the interval. For a new node joining the network, and requiring initialisation and therefore the establishment of synchronisation, a maximum slew rate of 100 ms/s was chosen, so as to ensure rapid convergence to the global time.

The feasibility of this implementation is seen from the results achieved in the experiments undertaken. Given the resulting synchronisation accuracy of less than 10 µs as shown in Fig. 3, it believed that such a solution is sufficient for many of the applications found within industrial control situations.

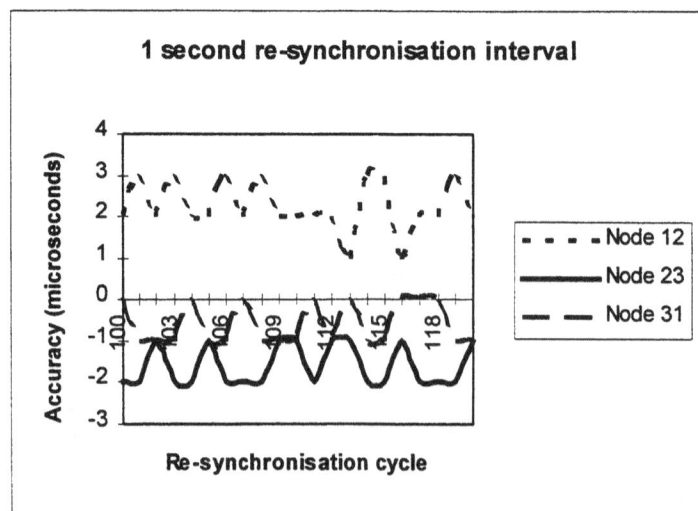

Fig. 3 Some experimental results

8

6. CONCLUSIONS

It has been suggested here that true real-time communication systems are an essential component of a real-time distributed computer control system. In addressing this requirement, this paper has emphasised the need for strict design and analysis procedures. To this end, it has been suggested that the whole design cycle must be supported by techniques and tools that ensure that the final products are both logically and temporally validated and verified. It has also been suggested that there are presently no such integrated tools available, and hence, a mixture of new and existing tools has to be carefully chosen. It has been proposed that a radically new temporal modelling tool, LIMITS, is appropriate, especially in that it provides a direct extension of the object modelling approach to design. LIMITS not only allows a full observation of all timing effects, but, as a formal approach to temporal modelling, allows a rigorous approach to design to be undertaken.

The points made in this paper have been applied to the development of a specific real-time fieldbus. This work began by investigating the CAN protocol and examining the fundamental necessity for providing a common time base across a CAN network. To achieve this, a hybrid approach to clock synchronisation was selected, as that seemed the most cost-effective solution, in terms of the accuracy achieved and the hardware required. This was proved in a laboratory environment to be a highly effective approach. The current implementation uses a 4-layer PCB plug-in card that provides any related application with simple, continuous access to a globally synchronised time. The feasibility of this approach can be seen from the results achieved in the experiments undertaken. Given the resulting synchronisation accuracy of less than 10 µs, it is believed that such a solution is sufficient to meet the needs of many industrial control applications.

REFERENCES

BONeS (1993), *BONeS Designer: Core Library Guide*. Comdisco Systems Inc., USA.

Burns, A. and Wellings, A. (1995), *HRT-HOOD™: A Structured Design Method for Hard Real-Time Ada Systems*. Elsevier, Oxford.

Dimyati, K. and Rodd, M.G. (1995), Real-time clock synchronisation for the CAN protocol. *Proc. of the 3rd. Baltic Summer School on Information Technology and System Engineering*, Kaunas University of Technology, Klaipeda, Lithuania.

Dimyati, K. (1996), *Real-time Clock Synchronisation over a CAN-based Fieldbus*. PhD thesis, University of Wales Swansea.

ISO 11898 (1993), *ISO 11898: Road Vehicles – Interchange of Digital Information – Controller Area Network (CAN) for High Speed Communication*. International Organisation for Standardisation (ISO).

Kopetz, H. (1997), *Real-time Systems – Design Principles for Distributed Embedded Applications*. Kluwer Academic Publishers, Boston, USA.

Lawrenz, W. (1995), World-wide status of CAN – Present and Future. *Proc. of the 2nd Intl. Conf. on CAN*, London.

Motus, L. and Rodd, M.G. (1994), *Timing Analysis of Real-time Software*. Elsevier Science/ Pergamon, Oxford, UK.

Motus, L. (1995), Timing problems and their handling at system integration. In S.G.Tsafestas and H.B.Verbruggen (eds), *Artificial Intelligence in Industrial Decision Making, Control, and Automation*, Kluwer Academic Publishers, 67-87.

Naks, T. (1996), *Towards Timed Object Modelling Techniques*. Engineering diploma thesis, Tallinn Technical University.

Olyeni, A. (1996), *Measuring Real-time Execution Time of Software on a PC using a Hardware Monitoring System*. Masters Thesis, University of Wales Swansea.

Quirk, W. J. and Gilbert, R. (1977), *The Formal Specification of Requirements of Complex Real-time Systems*. AERE, Harwell, No. 8602.

Reeve, A. *et al.* (1997), Feature on fieldbus. *Measurement and Control*, Vol. 30, No. 3, pp 68-81.

Rumbaugh, J., Blaha, M., Premerlani, W., Eddy, F. and Lorensen, W. (1991), *Object-oriented Modeling and Design*. Prentice Hall, Englewood Cliffs, NJ, USA.

TICKER (1994), *Ticker ASIC Hardware Specification*. Institute for Industrial Information Technology Limited (i^2it), Swansea, UK.

Törngren, M. (1995), *Modelling and Design of Distributed Real-time Control Applications*. PhD thesis, Dept. of Machine Design, Royal Institute of Technology, Sweden.

Welch, J. L. and Lynch, N. (1988), A new fault-tolerant algorithm for clock synchronisation. *Information and Computation*, Vol.77, pp. 1-36.

MODE HANDLING IN THE TIME-TRIGGERED ARCHITECTURE

H. Kopetz, R. Nossal, R. Hexel, A. Krüger, *
D. Millinger, R. Pallierer, C. Temple [*,1]

** Institut für Technische Informatik, Technische Universität
Wien, Austria*

M. Krug **

** Daimler Benz AG, Stuttgart, Germany

Abstract:
Large real-time applications often impose stringent requirements on the system architecture concerning the number of different operating conditions under which they must operate. Modes and mode changes are a powerful concept to represent these operational phases within the system architecture.
This paper focuses on the modes and the mode change strategy available in the time-triggered architecture. The time-triggered architecture is outlined and the principle of modes and mode changes is introduced. The modes introduced in the TTA meet the requirements imposed by different operating conditions of hard real-time systems. The requirements for mode changes are identified and the conflict between consistency and speed is resolved by introducing two distinct types of mode changes.
Keywords: Distributed Real-Time Systems, Static Scheduling, Modes

1. INTRODUCTION

Safety critical hard real-time application typically exhibit a number of mutual exclusive operational phases. Each of these phases requires different control activities. A computer system on board an aircraft, for instance, performs different actions during taxiing on the ground and during level flight with the autopilot engaged. Computer systems controlling hard real-time applications must offer real-time and fault-tolerance capabilities. The Time-Triggered Architecture (TTA) based on the Time-Triggered Protocol (TTP) (Kopetz and Grünsteidl, 1994) meets both demands.

The TTA, like any time-triggered system, relies on static scheduling of tasks and messages. These schedules are created pre-runtime, thus the system cannot adapt to changing conditions during runtime. This problem is overcome by introducing the concept of modes. A *mode* reflects a certain operational phase of the real-time application. Characteristics of a mode are the task set executed in this mode and the set of messages transmitted in this mode.

Mode changes are used to change between statically defined modes. They are the only means to change the temporal behaviour during operation. If the operating conditions or the operational phase change, the controlling computer system has to perform a mode change. Changing to another mode means to proceed according to a different temporal control pattern, i.e., a different task and message schedule. Work on

---

[1] This work has been supported in part by the ESPRIT project "Time-Triggered Architecture (TTA)" and the BRITE EURAM project "Safety Related Fault-Tolerant Systems in Vehicles (X-by-Wire)".

mode change extensions for a priority inheritance based scheduler is presented in (Sha *et al.*, 1989). Tindell et al. (Tindell *et al.*, 1992) deal with mode changes in a deadline monotonic scheduling algorithm. In the context of time-triggered architectures Fohler (Fohler, 1993) has developed a static scheduler capable of various modes and mode changes.

This paper shows how modes are integrated into the TTA and the protocol. The communication system of the TTA provides the mechanisms for mode handling to the application program. Hence the application has a powerful means for adapting to different conditions that is yet easy to use.

The rest of the paper is organized as follows. Section 2 gives an overview of the Time-Triggered Architecture focussing on time-triggered system activation and the data structures containing the temporal control pattern. An introduction to modes and mode changes is given in Section 3. Based on this general concept Section 4 goes into detail on the mode handling in the Time-Triggered Architecture. The paper is concluded in Section 5.

2. OVERVIEW OF THE ARCHITECTURE

In the Time-Triggered Architecture the controlling computer system consists of at least one *computational cluster* (figure 1). A computational cluster comprises a set of self-contained *node computers*. The nodes communicate via a (replicated) broadcast bus using the Time-Triggered Protocol. Therefore, a single cluster is also characterized as being a *broadcast domain*.

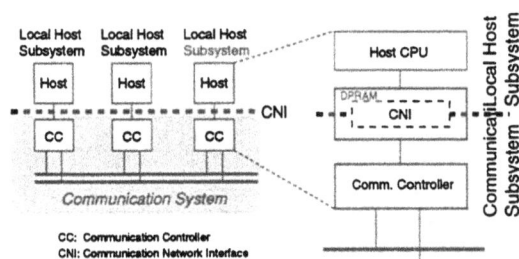

Fig. 1. Architecture of a Cluster and a Node.

Messages mutilated in the communication system are assumed to be detected and discarded. Further, nodes are assumed to be *fail-silent*, i.e., only crash failures and omission failures can occur. Self-checking mechanisms causing a high error detection coverage are employed to enforce this property. Extensive fault-injection experiments with the MARS prototype system (Kopetz *et al.*, 1989) have shown the feasibility of this assumption (Karlsson *et al.*, 1995).

Within a cluster, a global time base — of sufficiently small granularity with respect to a specific application — is established by synchronizing

the clocks located within the nodes (Kopetz *et al.*, 1995).

2.1 *Structure of a Node Computer*

Two main subsystems can be identified within a node computer: (1) the *communication subsystem*, which autonomously and reliably transmits messages between the nodes of the distributed system, and (2) the *host subsystem*, which executes the distributed real-time application. The operation of both subsystems is controlled by static configuration data structures. The *MEssage Descriptor List (MEDL)* resides in the communication subsystem (the TTP controller), and contains the attributes of the messages sent and received by the protocol. According to this list the TTP controller periodically and autonomously reads the messages to be sent from the CNI and writes received messages to the CNI. Similarly, the attributes of the tasks executed by the host processor under the control of a time-triggered operating system are stored in the *TAsk Descriptor List (TADL)*.

The interface between communication and host subsystem is called the *Communication Network Interface (CNI)* (Krüger and Kopetz, 1995). It is located in a dual-ported memory and enables the local host processors to send or receive messages by writing to or reading from this memory area. The main part of the CNI facilitating this message exchange is designed as a *data sharing interface* without any control information exchange. This reduces the probability of control error propagation and temporally encapsulates the subsystems. Consequently, the CNI can be seen as a temporal firewall between communication subsystem and host subsystem, each subsystem representing an error-containment region.

2.2 *Mechanisms of the Time-Triggered Protocol*

TTP is an integrated time-triggered protocol that provides prompt transmission of messages with high data efficiency, a responsive membership service, a fault-tolerant clock synchronization service, mode change support, error detection with short latency, and distributed redundancy management. TTP makes best use of the a priori information available in a TTA to reduce the number and size of messages, e.g., by retrieving the message identification from the a priori known time of message reception.

In TTP all nodes are forced to agree on their *controller states (C-states)*. The controller state consists of three fields: the MEDL position, the global time and the membership. The MEDL position field is a pointer to the current entry in the MEDL, i.e., it identifies the current mode and TDMA slot. The time field contains the global time at the beginning of the current TDMA slot. The membership field indicates which nodes have been active and inactive at their last membership point. To enforce C-state agreement between a sender and a receiver the CRC of a normal message is calculated over the message contents concatenated to the local C-state. A receiver can only interpret the frame if sender and receiver agree about the controller state at the time of sending and receiving. In case the C-state of the sender is different from the C-state of the receiver, any message sent by the sender will be discarded by the receiver, because the CRC check will fail.

Access to the transmission medium is controlled by a static TDMA scheme. Each node is allowed to send messages only during a predetermined time span, called its *TDMA slot*. The sequence of the periodic TDMA slots is called a *TDMA round*. With regard to the duration of the TDMA slots and to the sending sequence of the nodes, all TDMA rounds are equal. However, the length and contents (the application data) of the frames may differ. The set of periodically recurring TDMA rounds (with possibly different message length and contents) is called a *cluster cycle*.

To implement the fail-silence behavior assumed above, each node is equipped with two *bus guardians (BGs)*. These devices autonomously check and regulate the access times of a node to the bus, i.e., they grant bus access to the node only within its TDMA slots independent of the remaining protocol execution.

3. MODES AND MODE CHANGES

The time-triggered system paradigm as introduced in the previous section stipulates a constant behavior of the system, i.e., the temporal control pattern does not change over time. This is a rather inflexible approach that restricts the applicability of time-triggered real-time systems. Hence a mechanism to change the temporal control during operation has to be devised. In this section an appropriate mechanism, *modes* and *mode changes* will be introduced.

3.1 *Modes*

A possibility of classifying real-time applications is the dynamics with respect to their behavior. On the one hand there are applications whose behavior does not change during their operation. They continuously perform the same function[2].

An example for such a system is a hydro power plant. On the other hand there are applications that consist of a number of distinct phases. In each phase the system performs a certain function. Consider the example of an aircraft. A flight from point A to point B consists of the phases taxiing on the airport, takeoff, climbing, level flight, descending, landing, and taxiing again. In each of these phases some functions are important for the operation of the plane, whereas others are of minor priority or need not be carried out at all.

A mode of a real-time computer system reflects a certain operating condition or operational phase of the controlled object. Each mode of operation is characterized by a task set that is executed in this mode and a set of messages transmitted during this mode. Thus each mode has its unique temporal control structure, i.e., its own MEDL and its own TADL. Different modes have different task and message sets and a different temporal control pattern. However, the task and message sets need not be disjunct. For instance — revisiting the example given above — a task handling the gyro compass, a part of a plane's navigation system is needed during taxiing on the airport as well as during the plane's flight phase.

3.2 *Mode Changes*

Mode changes are the only means to change the temporal control structure during operation. In a mode the assigned, predetermined MEDL and TADL is executed in a periodic manner. This scheme does not allow for changes of the function of the system. If the operating conditions or the operational phase change, the controlling computer system has to perform a mode change. Changing to another mode means to proceed according to a different temporal control pattern.

Jahanian (Jahanian *et al.*, 1988) presented three types of mode changes according to the handling of the currently executing tasks.

(1) The easiest possibility is to abort all tasks and start a complete new task set in the destination mode. In this case the mode change can be performed immediately.
(2) The counterpart to this possibility is to complete all currently executing tasks and switch to the new mode then.
(3) Only some of the current activities are finished before the mode change takes place.

Fohler (Fohler, 1993; Fohler, 1994) advanced this concept and introduced so-called *Transition Modes*. The schedule of a transition mode is designed to appropriately handle the termination of activities.

---

[2] In this characterization maintenance or installation are not considered to be normal operational phases.

3.2.1. *Flexibility vs. Safety*

So far modes have been identified to be a powerful concept for changing a system's function during operation. Yet this flexibility introduces a number of potential dangers into the system.

Changes at Wrong Point in Time The most obvious danger is a mode change at a "wrong" point in time. A mode change occurs at a wrong point in time, if the state of the controlled object does not require a mode change. Considering the above example of an aircraft a mode change to taxiing mode would be wrongly timed if it occurred during level flight.

Inconsistencies Mode changes can cause inconsistencies between the states of the nodes of a cluster. If only a subset of the nodes perform a mode change two cliques of nodes are formed. Each clique assumes the communication pattern of the mode it is in, the cliques cannot communicate with each other. Thus, a correct operation of the system cannot be guaranteed any more.

3.2.2. *Requirements*

To avoid the problems arising a number of requirements for mode changes can be identified.

- The points in time when mode changes are allowed must be determined before runtime like the temporal control patterns of the modes. During runtime mechanisms must ensure that mode changes can only be performed at these preplanned points in time.
- It must be assured that mode changes are executed only when this is required by the state of the controlled object. Note that this requirement is different from the first. The first requirement determines the points in time when changes can be taken. As the system operates in a periodic manner, these points in time recur periodically. This requirement determines if a change is to be performed at one of these points.
- Concepts for guaranteeing the consistent change from a mode to another must be found. In this context consistent means that all nodes of a cluster change the mode at the same point in time. In other words all nodes of a cluster must agree on the current mode at all points in time.

4. MODE HANDLING IN TTA

The previous section presented the basic concepts of modes and mode changes. In this section the mode handling strategy of TTP is explained. It will be shown how mode changes are initiated and which restrictions apply. Two types of mode changes supported by TTP will be introduced, immediate and deferred mode changes.

4.1 *The Mode Change Process*

In the section "Modes and Mode Changes" it was stated that modes reflect certain operating conditions of the controlled object. Interaction with the environment is only performed by the host CPU. Thus only the host CPU knows about the operating conditions of the controlled object and only the host CPU is able to determine whether a mode change is required. Consequently mode changes in TTP are initiated by the host subsystem.

If the operating conditions of the controlled object change, the host CPU monitoring these conditions issues a mode change request to its local TTP controller. This implies that only certain nodes are able to initiate mode changes to certain modes.

4.1.1. *Performing the Mode Change*

In the following a brief overview of the mode change procedure is given. Explaining all mechanisms, especially the handling of failures, is beyond the scope of this paper.

Deferred mode changes	
Deferred change 1	010
Deferred change 2	100
Deferred change 3	110
Immediate mode changes	
Immediate change 1	011
Immediate change 2	101
Immediate change 3	111
Special meanings	
Stay in current mode	000
Clear pending deferred change request	001

Table 1. The meaning of mode change bits in the frame header.

After the host CPU has passed a mode change request to its local TTP controller (see above subsection), the latter checks whether the node is allowed to request this change. The information on allowed mode change requests is stored in the MEDL of the respective mode and is determined during the off-line planning phase of the system. If the requested change is allowed the TTP controller sets the mode change bits in the header of the next frame it sends to the value corresponding to the requested successor mode (see table 1). As there are three mode change bits in the frame header and two bit patterns are required to encode special actions ("stay in current mode" and "clear pending deferred mode change") six successor modes can be addressed.

All other nodes receiving the frame containing the mode change request in the header check in the current MEDL whether the sending node is allowed to request this mode change. If the requested change is allowed all nodes perform the mode change at the predetermined point in time (see following subsection).

It is important to note that the bus access pattern is identical in all modes, i.e., there is a fixed TDMA cycle pattern for all modes. This restriction is imposed by the Bus Guardian, in particular by its demand for utmost independence from the TTP controller. As mentioned in section "Overview of the Architecture" the Bus Guardian is not aware of the current operational mode of the computer system. Moreover, it is neither informed about mode changes nor is it able to notice them by himself. Thus different access patterns in different modes would interfere with the Bus Guardian's open-close mechanism.

4.1.2. *The Conflict between Consistency and Speed*

TTP allows up to six successor modes for each mode. The corresponding six mode changes are divided into two groups,

- three *deferred* mode changes and
- three *immediate* changes.

The two categories of changes reflect two different principles, consistency on the one hand and speed on the other hand. Whereas deferred mode changes are agreed upon by all nodes of the cluster and can be performed one TDMA cycle after the request at the earliest, immediate changes are executed immediately after the request. Both mode change concepts have specific advantages, neither of them is well suitable for all situations. Therefore TTP supports both principles, which are explained in more detail in the following subsections.

4.2 *Deferred Mode Changes*

Deferred mode changes can be used to reflect changes in normal operational conditions of the controlled object, e.g., a change from take-off mode to climb mode. As stated above deferred mode changes reflect the principle of consistency, i.e., maintaining an equal state among all nodes of a cluster and between the state of the protocol on the one hand and the state of the application on the other hand.

Deferred mode changes can be requested at any time during a cluster cycle. However, the mode change is performed at the beginning of the next cluster cycle. At this defined point in time the application should be in the *ground state* (Ahuja *et al.*, 1990). Application ground state means that no application task on any node is currently

executing (either running or preempted). For this reason no task must be aborted or finished in the succeeding mode.

The possibility to request a deferred mode change at any time during the cluster cycle implies that the mode change is pending till the start of the next cluster cycle. When this point in time arrives each node changes to the requested successor mode autonomously, i.e., without further notice. This automatism is applicable to all correctly operating nodes that have received the frame containing the mode change request. Nodes that reintegrate into the ensemble after this frame has been sent are not aware of the pending mode change. Hence they cannot change at the predetermined point in time. To overcome this problem pending mode changes become part of the C-State. An integrating node receiving an I-frame also receives information on the pending mode change and can perform the change consistently with the rest of the ensemble.

The deferred mode change mechanism also allows a form of agreement on mode changes. Using a special bit pattern in the frame header (see table 1) pending mode change requests can be cleared again.

The mechanism presented so far does not restrict requests in any way. This means that more than one request can be issued during a cluster cycle. In this case the deferred mode change requests must be prioritized. If more than one change is requested the highest priority change is executed. Prioritizing can be done either by assigning static priorities to the numbers of deferred mode change requests or by dynamically making the last request the highest priority one. In the first possibility deferred change request 1 could be assigned the highest priority and deferred request 3 the lowest.

Deferred mode changes are addressed relatively, i.e., the number of the requested mode change is used to look up in a table that contains the absolute mode number of the destination mode. As opposed to direct addressing, where three bits to encode a mode change request would result in only seven possible modes of the system, the indirect addressing scheme facilitates an unbounded number of modes.

4.3 *Immediate Mode Changes*

An immediate mode change is executed immediately at the end of the FTU slot where the mode change was requested. The change itself is not agreed upon and thus reflects the principle of speed.

The immediate execution of this type of mode change requests avoids two problems arising with deferred changes. First of all there is no need to prioritize changes. Since changes are carried out immediately after they have been requested there is at most one pending request at any time. For the same reason maintaining information about pending changes is not required.

Immediate mode changes are addressed absolutely, i.e., an immediate mode change leads to the same mode regardless of the mode the change was initiated from. Hence multiple immediate mode changes are idempotent. On the other hand the encoding scheme of the mode change bits in the frame header (see Table 1) restricts the number of immediate modes to three if direct addressing is used.

Though immediate changes can be handled easily in the communication subsystem they impose a number of serious problems on the application. Deferred changes are executed while the system is in ground state; immediate changes on the other hand cannot wait for such a ground state. Hence there are a number of application tasks executing when the mode change is invoked. These tasks must be aborted, which should not leave the application in an undefined state. In the destination mode the task set must be started amid a cycle unless there is no phase relation between the communication and the task schedule.

5. CONCLUSION

In this paper the concept of modes and mode changes as a means for changing the temporal properties of a time-triggered system has been presented. First, an overview of the time-triggered architecture was given. Then the basic concepts of modes and mode changes were outlined, followed by an explanation of the mode handling strategy used in the time-triggered protocol. Special emphasis was put on the description of the discrepancy between consistency and speed, which was addressed by the deferred and immediate mode change strategies employed in TTP.

The mode handling features presented in this paper are a service provided by the communication subsystem. A mode change strategy, making use of these mechanisms, must be implemented at the host level. Issues addressed by a mode change strategy encompass questions like how to reach agreement on the necessity of a mode change. In the context of the time-triggered operating system being developed by our group an appropriate mode change strategy will be defined and evaluated.

6. REFERENCES

Ahuja, M., A. D. Kshemkalyani and T. Carlson (1990). A basic unit of computation in distributed systems. In: *10th Int. Conf. on Distributed Computing Systems*. Paris, France. pp. 12–19.

Fohler, G. (1993). Realizing changes of operational modes with pre run-time scheduled hard real-time systems. In: *Responsive Computer Systems* (H. Kopetz and Y. Kakuda, Eds.). Vol. 7 of *Dependable Computing and Fault-Tolerant Systems*. pp. 287–300. Springer-Verlag.

Fohler, G. (1994). Flexibility in Statically Scheduled Real-Time Systems. PhD thesis. Technisch-Naturwissenschaftliche Fakultät, Technische Universität Wien. Wien, Österreich.

Jahanian, F., R. Lee and A. Mok (1988). Semantics of modechart in real time logic. In: *Proc. of the 21st Hawaii International Conference on Systems Sciences*. pp. 479–489.

Karlsson, J., P. Folkesson, Jean Arlat, Yves Crouzet and Günther Leber (1995). Integration and Comparison of Three Physical Fault Injection Techniques. In: *Predictably Dependable Computing Systems*. Chap. V: Fault Injection, pp. 309 – 329. Springer Verlag.

Kopetz, H., A. Damm, Ch. Koza, M. Mulazzani, W. Schwabl, Ch. Senft and R. Zainlinger (1989). Distributed Fault-Tolerant Real-Time Systems: The MARS Approach. *IEEE Micro* 9(1), 25–40.

Kopetz, H., A. Krüger, D. Millinger and A. Schedl (1995). A Synchronization Strategy for a Time-Triggered Multicluster Real-Time System. In: *Proc. 14th Symposium on Reliable Distributed Systems*. Bad Neuenahr, Germany.

Kopetz, H. and G. Grünsteidl (1994). TTP — A Protocol for Fault-Tolerant Real-Time Systems. *IEEE Computer* pp. 14–23.

Krüger, A. and H. Kopetz (1995). A Network Controller Interface for a Time-Triggered Protocol. In: *SAE Symposium on Future Transportation Electronics: Multiplexing and In-Vehicle Networking*. Society of Automotive Engineers. SAE Paper No. 952576.

Sha, L., R. Rajkumar, J. Lehoczky and K. Ramamritham (1989). Mode change protocols for priority-driven preemptive scheduling. *Real-Time Systems* 1(3), 243–265.

Tindell, K.W., A. Burns and A.J. Welling (1992). Mode changes in priority pre-emptively scheduled systems. In: *Proc. 13th Real-Time Systems Symposium*. pp. 100–109.

WORST-CASE EXECUTION TIME ANALYSIS AT
LOW COST

Peter Puschner [*,1]

* Technische Universität Wien, Austria

Abstract: The computation of Worst-Case Execution Times (WCETs) of tasks involves the description of possible execution paths on source level, the translation of the program and the path information to the machine level, the timing analysis at the machine level, and the documentation of the timing at the source level. These steps of WCET analysis strongly suggest the cooperation between compilers and machine-level WCET tools, and in fact some prototype compilers/WCET analyzers exist. To this date there exists, however, no commercial compiler that supports WCET analysis. This paper describes a light-weight approach to WCET analysis that does not depend on a specific compiler and thus reduces the dependency on compiler builders. The approach uses a tool to read the path information contained in the source code and map this information onto the assembler program. A low-level WCET analysis tool then computes the desired timing information of the annotated assembler program. Experiments with a simple prototype tool demonstrate the feasibility of the approach for simple programs and a moderately optimizing compiler.

Keywords: Worst-Case Execution Time Analysis, Execution Times, Performance Analysis, Real-Time Languages, Real-Time Computer Systems, Real-Time Tasks, Methodology

1. INTRODUCTION

In the past years worst-case execution time (WCET) analysis has become an acknowledged part in the theory of real-time systems construction. Researchers working on WCET analysis introduced extensions of high-level programming languages to describe possible execution paths (Kligerman and Stoyenko, 1986; Puschner and Koza, 1989; Park, 1993), constructed tools for machine-level WCET analysis (Harmon et al., 1992; Zhang et al., 1993; Li et al., 1995), and they built compiler prototypes to translate high-level source programs annotated with path information into code that is both executable and analyzable with respect to its WCET (Vrchoticky, 1994).

Despite the existence of the necessary research results in all these areas, there exists no commercial compiler on the market that translates both source code and information about possible execution paths during compilation, this way supporting WCET analysis. Real-time system programmers must thus still do without a WCET tool with a high-level language interface.

This paper proposes a light-weight approach to WCET analysis. The approach significantly reduces the dependency of WCET analysis from compiler builders and solves the above-mentioned problem. A simple prototype tool realizing the key parts of the approach has been implemented and evaluated. Instead of demanding from compilers that they translate path information from the source level to the machine level, the prototype tool uses knowledge about the program structure and line numbers of constructs to map between

[1] This work has been supported in part by the ESPRIT projects "Time-Triggered Architectures (TTA)" and "Design for Validation (DeVa)".

the two program representations. Having transferred the path information to the machine program, the WCET tool calls a WCET analyzer for assembler programs to compute detailed worst-case timing information (the latter was implemented in a student project at our department). Finally, the resulting timing information has to be mapped back to the source level.

The paper is structured as follows. Section 2 sketches the proposed approach for WCET analysis and lists the steps involved. Section 3 focuses on the problem of mapping the source-level path information given by the user to the assembler program generated by the compiler. It characterizes the representations of path information on both levels and discusses the key problem of mapping between the representations. Section 4 describes the experiences with a prototype tool that was implemented and Section 5 concludes the paper.

2. STEPS OF WCET ANALYSIS

Basically, a WCET tool that uses the proposed approach has to perform the following four steps to analyze a program (see Figure 1):

(1) *Structure and Path Analysis.* The WCET tool parses the high-level language source program annotated with path information — loop bounds, markers, frequency relations, and scopes (see below). From this input it builds a data structure that keeps information about the control structure, the line numbers of all control constructs, and the knowledge about the possible execution paths of the program.

In parallel the program is compiled with an *off-the-shelf compiler.* First, the source program is prepared for compilation. The path information constructs that a standard compiler cannot translate are either removed or translated into code for run-time checks. The program is then compiled with a standard, off-the-shelf compiler. The debug switches are turned on to produce debugging information with source line numbers with the translated code. (Most compilers can generate references to source line numbers, only the formats differ).

(2) *Mapping path information onto assembler code.* The result of the structure and path analysis and the assembler program generated by the compiler are used to map the path information from the original program to the machine-language program and annotate the assembler program with path information.

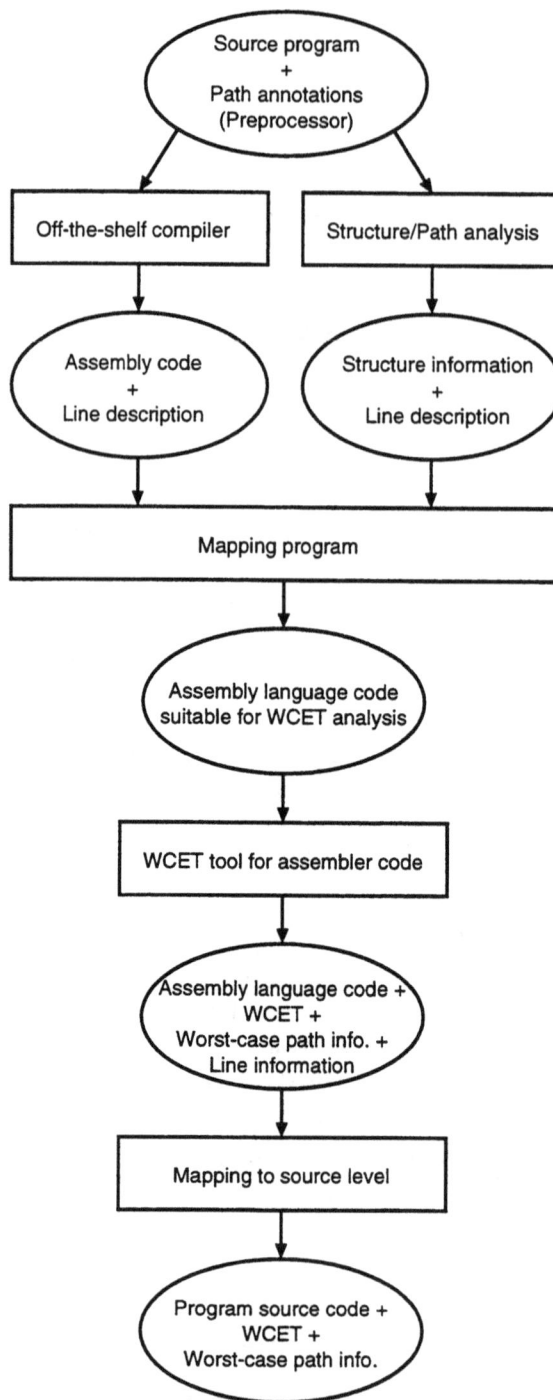

Fig. 1. Steps of WCET Analysis.

(3) *WCET Analysis.* The WCET of the machine-language program and the worst-case paths are computed with a WCET tool for assembler programs. This tool builds an integer linear programming problem for the computation of the WCET and calls an ILP solver to compute a solution. This solution consists of the WCET and information about execution frequencies of program parts in the worst case.

(4) *Documentation of Timing.* After the computation of the WCET, the information about constructs and line numbers is ac-

18

cessed again, to map the assembly-level timing information back to the source program.

The key problem in these four steps of WCET analysis is the translation of the path information from the source level to the machine-language level. The rest of this paper is dedicated to this problem.

3. PATH INFORMATION

This section shows how information about possible execution paths is represented both on source level and on machine level.

3.1 High-Level Path Description Constructs

The high-level path description constructs presented in sequel are language extensions for the C programming language which can be easily implemented as preprocessor statements. Note, however, that the language extensions can be implemented for any other procedural programming language with small effort.

3.1.1. Loop Bounds

A loop bound describes the maximum number of iterations of a loop in a given application. It is realized by the construct

$$\texttt{MAX\_ITERATIONS}(const\_expr).$$

The MAX_ITERATIONS statement must be the first statement in the loop body. It has one parameter, the loop bound, which must be a integral numerical constant expression.

Every loop must be bounded by a loop bound. Bounding all loops is, from the point of view of path description, sufficient to compute a bound for the WCET of a program. Restricting path information to loop bounds is, however, in general not sufficient for the computation of tight bounds.

3.1.2. Frequency Relations

Frequency relations are constructs that describe execution-frequency bounds of single program parts and to express disjunctive sets of linear relations between the execution frequencies of two or more program parts. Frequency relations implement the relative frequency constraints described in (Puschner and Schedl, to appear 1997).

Frequency relations are linear equations or inequalities. They characterize the execution frequencies of program parts marked with the *marker* construct:

$$\texttt{MARKER}(name)$$

The argument *name* is the unique name of the marker. In *frequency relations* marker names are used to describe the frequencies of marked program parts.

$$\texttt{RELATION}(f\_expr\ addop\ f\_expr\ addop\ \dots\ f\_expr \\ relop\quad const\_expr)$$

where a frequency expression *f_expr* either has the form

$$const\_expr * \texttt{FREQUENCY}(name)$$

or simply reads

$$\texttt{FREQUENCY}(name).$$

As before, *const_expr* represents an integral numerical constant expression. Further, *name* is the name of a marker that must have been defined before, *addop* stands for one of the additive operators '+' and '−', and *relop* stands for any relational operator, i.e., $relop \in \{\leq, <, =, >, \geq\}$. In frequency relations a FREQUENCY construct represents the execution frequency of the program part referenced by the respective marker name, and a frequency relation RELATION characterizes the execution frequency of one or more program parts in one execution of a program or *scope* (see below).

By default frequency relations are conjunctive. There is, however, also a need to express sets of frequency relations that exclude each other. The OR construct implements a disjunction between the frequency relations preceding it and the relations following it.

The constructs introduced so far are sufficient to characterize program paths. For programming convenience it is, however, desirable to limit the scope to which a set of frequency relations relates, i.e., to characterize the behavior of the program in a certain region of the code.

The keyword SCOPE in combination with a C block statement defines a scope. The scope starts at the begin of the block statement and ends at the end of the block statement. A scope contains a list of statements, including markers and other scopes, and a list of frequency relations that characterize the execution frequencies of the C statements contained in the scope for every execution of the scope.

The semantics of the frequency relations of a scope is that they characterize the execution frequencies of the marked program parts at the *end* of every execution of the scope. Note that frequency relations need not hold at *any* point in time during the execution of the scope.[2]

---

[2] Though some frequency relations are invariant (e.g., MARKER(A) <= 10), this is not true for the general case. E.g., the relation MARKER(A) - MARKER(B) == 0 might

19

```
SCOPE
{
    for(i=0; i<N; i++)
    {
        MAX_ITERATIONS(N);
        for(j=i; j<N; j++)
        {
            MAX_ITERATIONS(N);
            MARKER(M1);

            ... a[i][j] ... ;
        }
    }
    RELATION(FREQUENCY(M1) == N*(N+1)/2);
}
```

Fig. 2. Source Program with Path Annotations.

Figure 2 gives an example for the use of the introduced constructs. In this example the elements of the triangular matrix a are processed. The frequency relation describes that the program part that accesses the array elements is only executed $N \times (N+1)/2$ times, not N^2 times — the iteration limit the WCET tool would deduct if only the iteration limits of the loops were given.

3.2 Path Description at Assembly Language Level

At the assembly language level there exist no nested constructs. Except for subroutine calls and trap instructions an assembler program only consists of groups of instructions which are executed in sequence and points between instructions at which the possible execution paths fork or join. In accordance with the program structure, the set of path description constructs for assembly language programs is smaller than for source programs. Only markers and frequency relations exist. Markers are implemented as special comments in the form

|| marker *number*

where two vertical bars identify the begin of a special comment.[3] A positive integral number (*number*) following the *marker* keyword uniquely identifies the marked location in the program. Frequency relations look similar to the high-level relations:

|| relation *term relop term*

where *relop* is any relational operator and a *term* is constructed as follows

---

change its truth value several times during the execution of a scope before evaluating to true at the end of the scope.
[3] This syntax follows the notation for comments found in the MIT assembly language for the M68000 processor family. When working with other processors/assemblers, the comment tokens have to be adapted.

[*expr addop*]$^{+}$ *expr*

The operator *addop* represents an additive operator ($+$ or $-$) and an expression *expr* is either a frequency expression *f_expr* or an integral number. In assembly language programs, frequency expressions *f_expr* either have the form

number ∗ frequency *number*

or simply read

frequency *number*.

In both cases *number* stands for a positive integral number. The number following the *frequency* keyword stands for the number of a location that has been identified by a marker construct. As at the high level, groups of frequency relations may exclude each other. A special comment with the keyword OR implements a disjunction between the frequency relations immediately preceding it and the relations immediately following it.

3.3 Mapping the Representations

The difficulty of mapping the high-level representation to the assembly-language level can be summarized as follows. In the high-level language the program is well structured (assuming that no goto statements are used). It is composed of constructs with a single, clear defined entry point and, in most cases, a single point of exit. Only constructs that contain loop exits, return, or program-termination statements do not follow this structure. These jump statements can, however, be treated like jumps to the end of the loop or jumps to the end of the subroutine, respectively. They do thus not interfere with the control structure of other constructs.

Program translation does, in general, not maintain the simple program structure in which blocks with one entry and one exit can be characterized. On the contrary, compilers translate high-level constructs into pieces of assembler code with multiple entry and exit points. Also they may replicate code to make programs fast (see Figure 3).

The path matching program of the WCET analysis tool translates the path information provided in the source program into a corresponding path description on the assembly language level. It therefore has to find corresponding pieces of code in both representations, identify the possibilities for entering and leaving the translated constructs (e.g., loops and scopes), and insert appropriate path information into the assembler program.

The mapping of path information from the high level to the assembly-language level can be reduced to the translation of markers, scopes, and frequency relations. This is due to the fact that

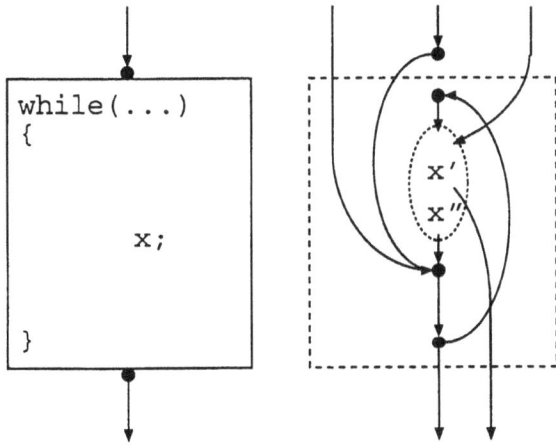

Fig. 3. Example of High-Level and Low-Level Representation of a Construct.

loop bounds are in truth only a special case of markers: The loop construct forms the scope for the marker. The marker is the first statement of the loop body and has associated a frequency relation describing the maximum number of iterations.

The following steps are necessary to map markers, scopes, and frequency relations to the assembler code:

- For each *marker* in the source code, search the corresponding program part in the assembler code and insert a marker there. Maintain a reference between the marker name in the source program and the number of the corresponding marker in the assembler program. Note that it can also be necessary to insert more than one marker into the assembler program for a single marker from the source (e.g., for a marker in a loop, if the loop is unrolled). In that case a reference to each of the inserted markers has to be stored.

- *Scopes:* In high-level code, scopes are the area of reference for frequency relations (relations are relative to the scope, e.g., to how often the scope is entered during program execution). Since the locality of frequency relations has to be conserved by the mapping, the mapping program generates reference points for the frequency relations of the translated program. It puts markers onto all paths that enter pieces of code resulting from the translation of the scope (see next point).

- *Frequency relations:* On the source level, frequency relations characterize the execution frequencies of markers relative to scopes. When translated to the assembler level a) the frequency relations of all markers have to be substituted by frequency expressions for their low-level counterparts and b) frequency relations have to be rewritten relative to the number of the possible entries into the scope.

4. EXPERIENCES

To investigate into the feasibility of the described approach, we built a prototype tool that performs the its core functionality. [4] In order to keep the efforts for tool construction as low as possible, the code of a WCET analysis tool for assembler programs that had been written in a former student project was reused. So, in consequence, only two components had to be programmed: the source-level parser for *structure and path analysis* and the *mapping program*, that transforms the high-level path information into annotations to the assembler program.

The parser for structure and path analysis reads source programs and builds a parse tree that contains information about the static structure of the programs together with the numbers of the lines at which constructs start and end, and the user-provided path information.

The mapping program tries to map the path information of the parse tree produced by the source-level parser onto the assembler program generated by the compiler. The mapping function of the first prototype is very simple. It assumes that programs are compiled with only moderate optimization (in the experiments the GNU GCC compiler was used at optimization level O1) and only generates single reference points for loops and scopes. The mapping function traverses the parse tree of the source program recursively. In each step down the tree it tries to find the begin and the end of the assembler code generated from the construct represented by the current node of the tree. It uses information about the nesting of constructs, line numbers, and information about jumps and loops in the assembler program to search for correspondences. In case of success the mapping program annotates the assembler program with path information generated from the source-level information. If the mapping function has problems to map a construct, it outputs a warning.

Tested with some small programs, the prototype tool analyzed most sample programs correctly. Only for some intentionally "malign" code samples the prototype could not map the path information successfully (e.g., in one example a single label worked as starting point for different loops; the tool could not insert a reference marker for the second loop). Of course, the simple mapping strategy does not cope with more sophisticated code optimization strategies (major re-arrangement of code, loop unrolling, etc.). In all cases that the tool could not solve, the tool reported the prob-

---

[4] Only the last step, which maps the results of WCET analysis back to the source level has not been implemented yet.

21

lems correctly, i.e., it did not produce wrong results without warning.

5. SUMMARY AND CONCLUSION

The paper presented a light-weight approach for the worst-case execution time analysis of real-time programs. Instead of constructing a compiler that transforms the information about possible execution paths described on the source level to the machine level, the described approach uses a mapping function that performs this transformation highly independent from the compiler used. After the mapping of path information, WCET bounds are computed with an existing tool for WCET analysis at the assembly-language level. The resulting detailed timing information of WCET analysis is then mapped back to the source level.

The results of experiments with a prototype tool that implements the core of our approach suggest that WCET analysis for high-level language programs is feasible, even if it is not supported by the compiler. Furthermore, the prototype implementation showed that the implementation of the proposed approach is far less expensive than the construction of a compiler that processes path information — the prototype could be developed within one man month.

An essential step of our future work will be an extension of the mapping function to cope with compiler optimizations.

6. ACKNOWLEDGEMENTS

The author is grateful to Emmerich Fuchs for his valuable comments on an earlier version of this paper.

7. REFERENCES

Harmon, M. G., T. P. Baker and D. B. Whalley (1992). A retargetable technique for predicting execution time. In: *Proc. 13th Real-Time Systems Symposium*. Phoenix, AZ, USA. pp. 68–77.

Kligerman, E. and A. Stoyenko (1986). Real-time euclid: A language for reliable real-time systems. *IEEE Transactions on Software Engineering* **SE-12**(9), 941–949.

Li, Y. S., S. Malik and A. Wolfe (1995). Efficient microarchitecture modeling and path analysis for real-time software. In: *Proc. 16th Real-Time Systems Symposium*. pp. 298–307.

Park, C. Y. (1993). Predicting program execution times by analyzing static and dynamic program paths. *Real-Time Systems* **5**(1), 31–62.

Puschner, P. and Ch. Koza (1989). Calculating the maximum execution time of real-time programs. *Real-Time Systems* **1**(2), 159–176.

Puschner, P. and A. Schedl (to appear 1997). Computing maximum task execution times — a graph-based approach. *Real-Time Systems*.

Vrchoticky, A. (1994). Compilation support for fine-grained execution time analysis. In: *Proceedings of the ACM SIGPLAN Workshop on Language, Compiler and Tool Support for Real-Time Systems*. Orlando FL.

Zhang, N., A. Burns and M. Nicholson (1993). Pipelined processors and worst case execution times. *Real-Time Systems* **5**(4), 319–343.

END-TO-END DESIGN OF DISTRIBUTED REAL-TIME SYSTEMS

Minsoo Ryu and Seongsoo Hong

School of Electrical Eng. and ERC-ACI
Seoul National University
Seoul, 151-742, Korea
{msryu, sshong}@redwood.snu.ac.kr

Abstract: This paper presents a systematic approach to translating high-level end-to-end timing constraints into a set of task-level timing constraints on a complex distributed real-time control system. This problem is formulated as an optimization problem so that the system design goals can be easily incorporated into the problem and the entire process can be automated. This formulation allows for an automatic tool-based approach, thereby allowing rapid prototyping of designs and identification of system design bottlenecks.

Keywords: System Design, Optimization, Timing

1. INTRODUCTION

As demands for factory automation increase, industrial distributed control systems are required to integrate a large number of heterogeneous control computers. Such heterogeneity and complexity require that industrial distributed control systems should be designed and implemented in a composable fashion such that a system is built by merely integrating constituent components.

Unfortunately, it is extremely difficult to design a distributed control system in a completely composable fashion in the presence of timing constraints. The reason is that the temporal relationships induced by the timing constraints may introduce coupling between structurally irrelevant components. A possible solution to this problem is to map system-level timing constraints onto component-level timing constraints based on the functional structure of the system. In many real-time system design methodologies, a system component often corresponds to a *task* and system-level *end-to-end* timing constraints postulate on the external inputs and outputs of the system (Gomaa, 1984; Gerber *et al.*, 1995). Thus, if there exists a mapping from end-to-end constraints onto task-level constraints, the composability of distributed real-time system design can be achieved.

Figure 1 shows an overview of the methodology of interest. A system design specified as a task graph and end-to-end constraints on the inputs and outputs of the system are given as input. Also given as input is the result of allocating the tasks to the host processors for a given distributed platform. This is denoted as task allocation in Figure 1. Based on these inputs task specific attributes such as task periods and deadlines are automatically derived. These results are used to schedule the tasks, and finally an executable system is generated to be deployed onto the given target distributed platform.

In this paper, the focus is on the derivation of task specific attributes in the design methodology. The problems of real-time scheduling have been well addressed in the literature (Burns, 1994). The

The work reported in this paper was supported in part by Engineering Research Center for Advanced Control and Instrumentation (ERC-ACI) under Grant 96K3-0707-02-06-1, KOSEF under Grant 96-2037, S.N.U. Korea Electric Power Corp. Research Fund under Grant 96-15-1135, and NSERC Operating Grant OGP0170345.

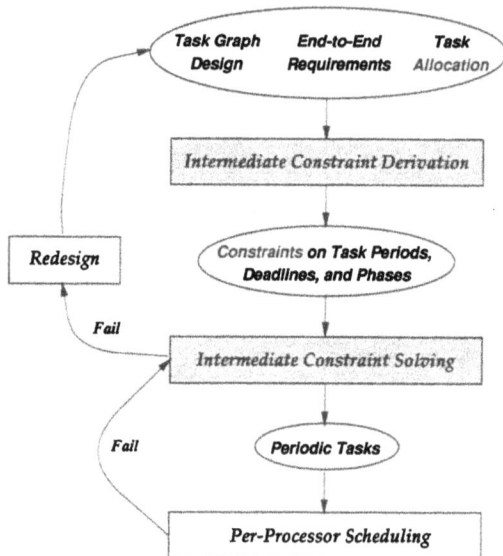

Fig. 1. The structure of the end-to-end design methodology.

main contribution of this work is in the development of a systematic methodology to transform a high-level design into a schedulable system.

2. THE MODELS

The system design model is defined as a task graph and associated end-to-end timing requirements while the implementation model possesses a set of periodic tasks and task specific attributes.

2.1 The Target Platform

In many distributed real-time control systems, sensors are polled periodically for external inputs which are then processed by one or more controller tasks to generate commands for the actuators. While simple control systems may consist of a single such control loop, systems which possess many interacting control loops executing at different rates sharing tasks and data are becoming more prevalent. Furthermore, in these systems the control loops may be executing concurrently while physically distributed over a number of processing hosts.

Figure 2 shows a typical configuration for a distributed real-time control system. The system has two layers which are connected through a gateway station. The application layer is a non-real-time environment where the computer systems are generally used for supervisory purposes. The field layer is the layer that requires stringent timing constraints. The sensors and actuators are autonomous devices connected to the processing hosts via a real-time communication medium such as a Controller Area Network (CAN) (Tindell

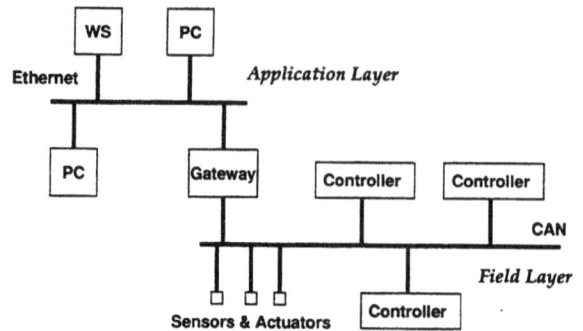

Fig. 2. Layered architecture of a distributed control system.

et al., 1994) or a FieldBus (Pleinevaux and Decotignie, 1988). Each processing host at the field layer is a single CPU controller and all hosts are assumed to be synchronized with respect to a global time base. The eventual goal, then, of the methodology presented in this paper is in designing the field layer.

For this paper, two assumptions are made. First, each processing host must have a suitable real-time operating system which is used to execute periodic real-time tasks. Second, the communication network is capable of guaranteeing bounded message transfers. Note that there exists numerous analytic methods to bound the worst case communication delay of periodic real-time messages (Tindell *et al.*, 1994; Pleinevaux and Decotignie, 1988).

2.2 Design Model

2.2.1. Task Graph.

A simple task graph model defined as a directed graph $G(V, E)$ is used to represent a real-time system, where

- $V = \mathcal{T} \cup \mathcal{P}$ is a set of nodes where $\mathcal{T} = \{\tau_1, \ldots, \tau_n\}$ is a set of tasks and $\mathcal{P} = \{\pi_1, \ldots, \pi_m\}$ is a set of communicating ports,
- $E \subset (\mathcal{T} \times \mathcal{P}) \cup (\mathcal{P} \times \mathcal{T})$ is a set of directed edges between tasks and ports; $\tau_i \to \pi_j$ denotes task τ_i's write access to port π_j and $\pi_j \to \tau_i$ is τ_i's read access to π_j.

Figure 3 gives an example of such a task graph. The properties of the task graph model are as follows:

(1) Tasks perform the computational activities in the system which consists of reading data from all its input ports, operating on the data, and finally, writing data to its output ports.

(2) Tasks communicate with each other through ports. The ports are accessed by the tasks through generic operations such as "Read" and "Write." Writes to a port are always

asynchronous and non-blocking. On the other hand, reads are synchronous, that is a process reading from a port waits for data to be written.

A real-time system designed with this model is thus represented as a finite, directed, acyclic graph where the tasks and the ports form the nodes in the graph and the edges correspond to reads and writes to the ports. In such a design the path from a sensor to an actuator forms a chain of producer/consumer pairs forming an end-to-end computation. Timing constraints are often defined on such end-to-end computations. Let $\Gamma(\mathcal{S}\|\mathcal{A})$ represent a transaction that takes inputs from sensors in set $\mathcal{S}$ and produces outputs for actuators in set $\mathcal{A}$. The transaction then consists of all the tasks that fall on the path from any sensor in $\mathcal{S}$ to any actuator in $\mathcal{A}$.

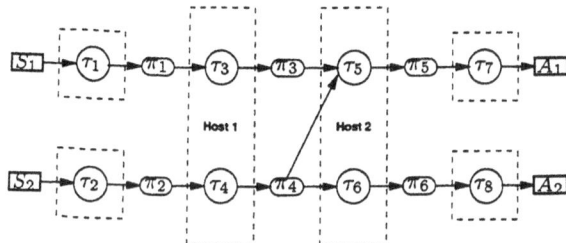

Fig. 3. Task graph design with task allocation.

2.2.2. *Task Allocation.*

Structurally, removing all nodes that correspond to network ports and all their incident edges in a given task graph results in a set of disjoint subgraphs. These disjoint subgraphs are referred to as *partitions*. In Figure 3, $\langle \tau_3, \tau_4 \rangle$ and $\langle \tau_5, \tau_6 \rangle$ denote two such partitions. Each partition is then mapped to a host. The mapping of partitions to hosts defines the allocation of tasks. For this paper, this is a predefined static mapping that does not change during run-time.

2.2.3. *End-to-End Timing Constraints.*

The following types of timing constraints are established on the transactions.

(1) *Maximum Allowable Validity Time (MAVT)*: This constraint bounds the maximum end-to-end delay permissible between the reading of a sensor and the delivery of the output command to an actuator based on that reading. Notation $M(S\|A)$ is used to represent the maximum delay from sensor S to actuator A.

(2) *Input Data Synchronization*: This constraint ensures that when multiple sensors collaborate in driving an actuator, then the maximum time-skew between the sensor readings is bounded. Notation $Sync(S_1, S_2\|A)$ is used

to denote the maximum time-skew between two sensors S_1 and S_2.

(3) *Maximum Transaction Period (MTP)*: This constraint bounds the maximum activation period for an end-to-end computation. Notation $MaxP(S\|A)$ is used to denote the maximum activation period.

2.3 *Implementation Model*

The real-time system of interest is assumed to be implemented as a set of periodic tasks. A periodic task τ_i, then, is represented by a 5-tuple $\langle e_i, T_i, d_i, \phi_i, \mathcal{P}_i \rangle$ where e_i represents the task's execution delay, T_i its activation period, d_i its deadline relative to the start of period, ϕ_i its initial phase (denoting the initial activation time of the periodic task), and $\mathcal{P}_i$ the processor to which it is allocated. The message stream m_i is treated analogously with a task with notation $\langle e_i^m, T_i^m, d_i^m, \phi_i^m, \mathcal{P}_i^m \rangle$.

3. PROBLEM DESCRIPTION AND SOLUTION OVERVIEW

Having described the model, in this section, a precise description of the problem is given along with an overview of the solution.

Given a task graph design of a system with the task allocation (Figure 3) and the end-to-end timing requirements the problem now is to derive task specific attributes (i.e., the activation period, deadline, and phase) for each task and message stream. This problem is modeled as a constraint solving problem in which the end-to-end requirements are first expressed as a set of intermediate constraints on the task attributes.

The next step is to solve the intermediate constraints such that the results preserve the timing correctness, i.e., if the final task set (which is a solution of the constraints) is schedulable, then the original end-to-end requirements will be satisfied. In so doing, arbitrarily complex schedulability analyses of the final task set is not addressed in this paper. However, a utilization-based schedulability analysis is incorporated into the constraint solving process. This schedulability analysis, though simple, makes it possible to avoid solutions that are trivially unschedulable. Generally, however, solving such constraints is not an easy problem due to several factors. First, in a distributed system there may be many tasks, and that may induce many variables. Second, the intermediate constraints are not always linear, as will become clear in Section 4. Third, incorporating schedulability into the constraint solving process significantly complicates the problem.

One way to tackle all these complexities is to use optimization techniques such as genetic algorithms or simulated annealing (Tindell *et al.*, 1992). These approaches, however, have a major common drawback in that they do not give any feedback when a solution cannot be found. To overcome this problem, the approach taken in our methodology is to decompose the constraint solving problem into a sequence of sub-problems. The motivation behind this approach is that specialized heuristics and performance metrics may be employed for each sub-problem. This, in turn, will allow for easier analysis of the system, thereby providing useful feedback upon failure. However, the success of the overall approach depends critically on whether the constraint solving problem can be decomposed into well-defined sub-problems. For this paper, this problem is decomposed into the *Period Assignment* and the *Phase and Deadline Assignment* sub-problems which are outlined below.

- *Period Assignment:* The first sub-problem assigns activation periods to tasks with an objective of minimizing the utilization of processors. Since the utilization depends only on periods and execution times of tasks, it can be computed without having knowledge of any other task attribute.
- *Phase and Deadline Assignment.* This sub-problem determines the phase and deadlines of each task based on the periods determined in the first sub-problem. The main problem here is to determine a set of individual task deadlines such that it is feasible to schedule each task within its deadline. The phases are determined last, and any solution consistent with the constraints is acceptable. As phases are used to maintain only precedence among tasks and input data synchronization, no schedulability measure is associated with them.

It is essential that the two sub-problems are solved in the given order as it is difficulty to obtain any useful schedulability criteria the other way around. Precise specifications of the sub-problems are given after showing how the entire constraint problem is setup.

4. DERIVING THE INTERMEDIATE CONSTRAINTS

The first step in the methodology is to transform the end-to-end constraints and synchronization requirements into a set of intermediate constraints on task attributes.

4.1 *Intermediate Constraints from the Producer /Consumer Model*

The producer/consumer model forms the basic communication semantics in our transaction model. A producer/consumer pair inherently incurs blocking synchronization in that a consumer task must wait for a producer task to generate the needed data. Such synchronization can be satisfied when the period of the consumer is an integral multiple of that of the producer. This relation is referred to as being *harmonic*. Therefore, to enforce this synchronization, a *harmonicity* constraint is imposed on all producer/consumer pairs. This is represented as $T_p | T_c$, where the operator "|" is interpreted as "exactly divides."

4.2 *Intermediate Constraints from Precedence and End-to-End Delay*

One of the end-to-end properties of interest is the end-to-end delay from a sensor reading to an actuator output. Since data flow through the tasks on the path from the sensor to the actuator and delay is incurred at each step, computation of the intermediate delays is required. This flow of data is a producer/consumer model, and recall that under producer/consumer synchronization a precedence relation between the producer and consumer is required. In real-time systems, an attractive way to enforce this precedence is through the use of task phase variables. Thus, for a given producer/consumer pair (τ_p, τ_c) and message task τ_p^m, the following intermediate constraints due to the precedence constraints are enforced.

$$\phi_p^m \geq \phi_p + d_p \quad \text{and} \quad \phi_c \geq \phi_p^m + d_p^m \quad (1)$$

Note that d_p^m is constant due to the aforementioned assumption, and that these precedence constraints must be satisfied for each producer/consumer pair.

The next intermediate constraint has to do with the end-to-end delay, that is, the maximum delay from the sensor to the actuator. The maximum delay from the time τ_p reads data from its input ports to the time τ_c writes data to its output ports is when τ_p reads its input data just as it is invoked and τ_c finishes writing to the output ports at the end of its activation period. Therefore, the worst case delay becomes $\phi_c - \phi_p + d_c$. By extending the same logic to a transaction, the following end-to-end intermediate constraint must be satisfied.

$$\phi_a - \phi_s + d_a \leq M(S \| A) \quad (2)$$

4.3 Intermediate Constraints from Input Data Synchronization

A variety of sensing devices such as switches, counters, and analog to digital (A/D) converters are used in the field layer. Each of these devices usually consists of two functional modules: a hardware module executing the virtual sensor task and a network adaptor module executing the virtual message task. The sensor task (producer) produces the sensor readings which are delivered to the control tasks (consumers) by the message task.

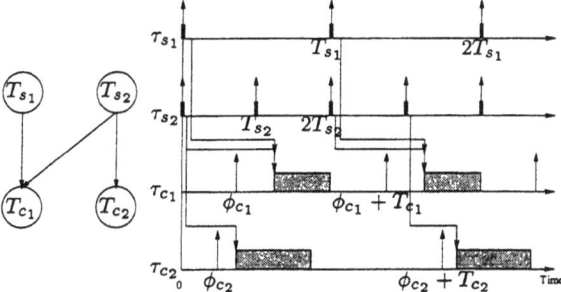

Fig. 4. Input data synchronization between two sensors.

Now, consider the input data synchronization requirement $Sync(S_1, S_2 \| C_1)$ where the sensor inputs S_1 and S_2 must be synchronized. Let τ_{s_1} and τ_{s_2} be the sensor task reading from S_1 and S_2, respectively. Then, the corresponding message tasks are $\tau_{s_1}^m$ and $\tau_{s_2}^m$. Figure 4 depicts an example of the sensor data synchronization and the communication required between the sensors and the consumers. Note that due to the assumption that the execution time of the sensor task is $e_s = 0$ the deadline of the sensor task is $d_s = 0$. This means that a sensor reading is available at the beginning of every activation period of the sensor task. Since the activation of the sensor tasks should be made at the beginning of any transaction, the initial phases of all the sensors should be equally set to zero. The message tasks of the sensors also start at the activation time of the sensors as shown in Figure 4. That is,

$$\phi_{s_1} = \phi_{s_1}^m = \phi_{s_2} = \phi_{s_2}^m = 0.$$

Referring to the two sensor example in Figure 4 the following harmonicity constraints may be imposed between the producer and the consumer.

$$T_{s_1}^m | T_{c_1}, \quad T_{s_2}^m | T_{c_1}, \quad \text{and} \quad T_{s_2}^m | T_{c_2}$$

Then, from the harmonicity constraints, T_{c_1} is the least common multiple (LCM) of $T_{s_1}^m$ and $T_{s_2}^m$, and thus consumer task τ_{c_1} uses the sensor readings which are sampled at every LCM times of $T_{s_1}^m$ and $T_{s_2}^m$. This shows that the input data synchronization requirement is satisfied if the harmonicity constraints imposed between the producers (sensors) and consumers are satisfied.

5. SOLVING THE INTERMEDIATE CONSTRAINTS

Once the intermediate constraints are derived, they are solved for task attributes such as task periods, deadlines, and phasings. The set of intermediate constraints and associated objective functions formally define the problem to be solved for the final result. In this approach, utilization on each computing host is chosen as the objective function.

5.1 Period Assignment

The first step in the constraint solving process is the assignment of periods to tasks. The constraints on task periods arise from two sources. They are

- *Range Constraints:* These constraints ensure a minimum rate of execution on tasks and therefore, impose an upper-bound on the periods. An implicit lower bound exists for every period variable which arises from its execution time.
- *Harmonicity Constraints:* These constraints arise from synchronous communication between each producer/consumer task pair.

Recall that for this paper, minimizing the processor utilization is the objective function of choice. More formally, for a set $\mathcal{T}^i$ of tasks allocated on processor $\mathcal{P}^i$, the utilization of processor $\mathcal{P}^i$ is defined as

$$U^i = \sum_{\tau_i \in \mathcal{T}^i} \frac{e_i}{T_i}$$

where $\mathcal{T}^i = \{\tau_1, \tau_2, \ldots, \tau_m\}$ and T_i is the period assigned τ_i. Then, the period assignment problem is to find a set of periods, T_i, such that (1) the above constraints are satisfied, and (2) $\max U^i$ is minimized. The rationale behind condition (2) is that in a schedulable system no processor should be overloaded. Therefore, this is simply an attempt to minimize the overall utilization on the most heavily loaded processor.

The solution strategy taken is to solve this problem in two stages: first, the feasible solution space is reduced as much as possible, and then, a branch-and-bound search is performed on the reduced search space (Gerber *et al.*, 1995).

5.2 Phase and Deadline Assignment

Once the periods are known, solutions for phase and deadline variables are derived. This process is done in several steps, as outlined below.

(1) *Phase Variable Elimination.* Using Fourier Variable Elimination (Dantzig and Eaves, 1973) the phase variables are eliminated from the precedence and delay constraints.

(2) *Deadline Decomposition.* The phase variable elimination step yields a system of constraints on the deadlines. Suppose the ith constraint in the constraint system has n_i deadline variables. Then, the ith constraint is of the following form.

$$d_{i_1} + d_{i_2} + \cdots + d_{i_{n_i}} \leq D_i$$

Thus, deriving each of the intermediate deadlines involves decomposing the end-to-end deadline D_i. The following simple heuristic is used for deadline decomposition.

Assign intermediate deadlines proportional to the task's period. Thus,

$$d_{i_k} = \frac{T_{i_k}}{\sum_{j=1}^{n_i} T_{i_j}} \cdot D_i$$

When there are several constraints on d_{i_k}, d_{i_k} may have multiple values. Then, the smallest d_{i_k} is taken as the deadline.

This simple heuristic results in a task set whose deadline monotonic priority order is consistent with the task precedence order in the transaction.

(3) *Phase Assignment.* Finally, once the deadlines have been assigned, the values for task phases are determined. For each phase variable, the smallest value which satisfies the constraints is assigned.

6. CONCLUSION

There are two directions along which this approach can be extended. First, as proof-of-concept, implementation of this methodology as a full-fledged software package is in progress. A single processor version has already been implemented, and extensions to a distributed version is in progress. The other direction is in extending our design model to incorporate more elaborate task structures, communication mechanisms, and timing constraints.

7. REFERENCES

Burns, A. (1994). Preemptive priority based scheduling: An appropriate engineering approach. In: *Principles of Real-Time Systems* (S. Son, Ed.). Prentice Hall.

Dantzig, G. and B. Eaves (1973). Fourier-Motzkin Elimination and its Dual. *Journal of Combinatorial Theory (A)* **14**, 288–297.

Gerber, R., S. Hong and M. Saksena (1995). Guaranteeing real-time requirements with resource-based calibration of periodic processes. *IEEE Transactions on Software Engineering* **21**(7), 579–592.

Gomaa, H. (1984). A software design method for real-time systems. *Communication of the ACM* **8**(2), 938–949.

Pleinevaux, P. and J. D. Decotignie (1988). Time critical communication networks: Field buses. *IEEE Network Magazine*.

Tindell, K., A. Burns and A. Wellings (1992). Allocating real-time tasks (an NP-hard problem made easy). *The Journal of Real-Time Systems* **4**(2), 145–165.

Tindell, K., H. Hansson and A. Wellings (1994). Analysing real-time communications: Controller area network (CAN). In: *Proceedings, IEEE Real-Time Systems Symposium*.

A SEMANTICS-PRESERVING TRANSFORMATION
OF STATECHARTS TO FNLOG

Arcot Sowmya* S. Ramesh**

*School of Computer Science and Engineering, University of New
South Wales, Sydney, NSW 2052, Australia
**Department of Computer Science and Engineering, Indian
Institute of Technology, Powai, Bombay 400 076, India

Abstract: Statecharts is a formal visual specification language for complex discrete-event entities. While the language is excellent for specifying real-time reactive systems and to build simulations and prototypes rapidly, statecharts suffers from the fact that the specifications are not verifiable. To fulfill the need for verification, we have earlier designed a logical specification language called FNLOG, which permits verification of statecharts. Verification is facilitated by a a semantic bridge between statecharts and FNLOG, such that for a system being specified, both its statecharts specification and its FNLOG specification map onto the same semantic domain of histories. This paper describes the transformation rules for transforming statechart specifications into FNLOG specifications and shows that the process is semantics-preserving. It then outlines a proof system for statecharts based on a verification of FNLOG specifications.

Keywords: Real-time systems, Statecharts, Temporal logic, Transformational approach, Verification.

1. INTRODUCTION

Statecharts is a formal visual specification language for complex discrete-event entities. It is an extension of the state/event formalism which satisfies software engineering principles such as structuredness and refinement, while retaining the visual appeal of state disgrams (Harel, 1987). The language was designed specifically for real-time reactive systems, which interact continuously with their environment and react dynamically to changes in them, with the interactions driven purely by events inside and outside the system. Such systems usually exhibit concurrency properties. Examples include embedded control systems, telecommunication systems and man-machine interfaces for software systems. Due to its visual nature, statecharts specifications are easy to develop, while its formal syntax and semantics enable quick development of simulations and executable specifications. Any level of detail is permitted, so that entire system behaviour may be specified. However, statecharts suffers from the fact that specifications are not verifiable at all, even though formal. Specifications can get loaded with detail, with many nested levels which can get difficult to check for correctness.

To fulfill the need for verification, we have earlier designed a logical specification language called FNLOG, which permits verification of statecharts (Sowmya and Ramesh, 1992). FNLOG is an independent specification language based on first-order predicate logic with arithmetic, extended by quantified temporal operators; since it is logic-based, it possesses a deductive proof system so that FNLOG specifications are verifiable. How does FNLOG aid in the verification of statecharts? To facilitate it, we have built a semantic bridge between statecharts and FNLOG, such that for a system being specified, both its statecharts specification and its FNLOG specification map onto the same semantic domain of histories (Sowmya and Ramesh, 1994). In designing this mapping, we have exploited the compositionality of statecharts, by mapping syntactic components of statecharts onto sets of FNLOG formulae. For FNLOG itself, we define a semantic mapping which maps elements of FNLOG onto the set of histories of a corresponding statechart. Then, the mapping from statecharts components to FNLOG formulae preserves semantics, in the sense that both specifications will refer to the same subset of histories.

The semantics-preserving mapping from statecharts to FNLOG enables the automatic derivation of an equivalent FNLOG specification from a given statecharts specification. Since system properties may be easily stated in FNLOG, they may then be verified against the FNLOG specification using its proof system. Since the FNLOG specification is an exact image of the original statecharts specification, the property then is true of the statecharts specification too. The proof system for FNLOG is based on the deductive proof system for first order predicate logic. The axioms and proof rules are augmented by special axioms which exploit the elements of FNLOG.

While the transformations from statecharts components to FNLOG specifications have been informally alluded to in earlier papers, they have never been formally presented. This paper describes the transformation of statechart specifications to FNLOG specifications and shows that it is semantics-preserving. It then outlines a proof system for statecharts based on a verification of FNLOG specifications. Section 2 briefly recapitulates the compositional syntax and history-based semantics of statecharts and section 3 presents an overview of FNLOG. Section 4 presents the mappings from statecharts components to FNLOG specifications and their semantic equivalence. Section 5 outlines a proof system for this scheme and section 6 concludes the paper.

2. COMPOSITIONAL SYNTAX AND SEMANTICS OF STATECHARTS

We use a non-graphical syntax of statecharts proposed by Hooman et al. (Hooman *et al.*, 1989) according to which any statechart is built up from primitive objects and some operators, which have a natural relationship with the pictures. The primitive objects are the so-called Basic statecharts [I,O,S] where S is a state name, I a set of incoming arcs and O a set of outgoing arcs. Only the outgoing arcs are labelled with an "event [condition] / action".

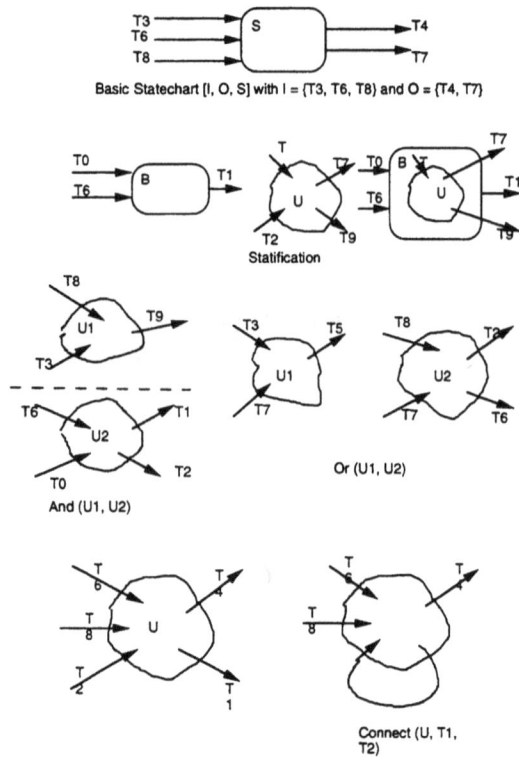

Basic Statechart [I, O, S] with I = {T3, T6, T8} and O = {T4, T7}

Statification

And (U1, U2)

Or (U1, U2)

Connect (U, T1, T2)

Fig 1. Syntactic Components of Statecharts

Let B be a basic statechart. U's are statecharts, T's are transition names and e is an atomic event. The operators that operate on statecharts are defined below. The operations are illustrated in Fig. 1.

(1) *Statification*: Stat(B, U, T) makes the state B a superstate with U inside it, with the incoming transition T of U as its default entry.

(2) *OR-construct*: Or (U_1, U_2) leads to a statechart which becomes an OR-state after statification.

(3) *And-construct*: And (U_1, U_2) yields an AND-state after statification.

(4) *Connect*: Connect (U, T_1, T_2) results in a statechart identical to U except that outgoing arc T_1 and incoming arc T_2 of U are connected to form a single complete transition.

(5) *Hide-closure*: HiCl (U, e) hides any generation of e by U (hiding) and makes U insensitive to any e generated by the environment (closure).

The semantic model associates with a statechart the set of all maximal computation histories representing complete computations. The semantics is a "not-always" semantics in which transitions labelled with $\neg e/e$ will never trigger, so that deadlock eventuates. It has been shown in (Huizing *et al.*, 1988) that, besides denotations for events generated at each computation step (the observables) and denotations for entry and exit, the following is necessary and sufficient to obtain a compositional semantics:

(1) a set of all events assumed to be generated by the whole system at each step, including the environment and

(2) a causality relation between the generated events.

A computation history h of a statechart U is of the form $h = (\hat{s}, i, f, o, s)$ where

- $\hat{s}$ models the start step,
- i is an incoming transition, or * to model implicit entry
- $f : N \rightarrow \{ (F, C, <, P) \mid F \subset C \subset E_e , < $ a total order on C $\}$ records for every step n a triple $(F, C, <)$ where
  - N is the set of natural numbers
  - F is a subset of the events generated by U. Considering the chain of transitions in step n, F contains the events which are generated by U, for the first time in the chain
  - C is a set of events generated by the total system in step n, including F
  - $<$ denotes the causal relationship between events generated by the whole system. If a causes b, then $a < b$.
  - P is an n-ary set of values for the variables; P could contain n-tuples of values for n variables.
- o is an outgoing transition, or * for an implicit exit, or $\perp$ when there is no exit
- $s \in N \cup \{ \infty \}$ denotes the exit step.

For a function f as defined above, the fields of f (n) are selected by f^F (n), f^C (n) and $f^<$ (n).

Define $\mathcal{H} = \{ h \mid \hat{s} < s, o = \perp \leftrightarrow s = \infty,$ and $(v \leq \hat{s} \vee v > s) \rightarrow f^F (v) = \phi \}$

Our semantic domain is given by $(D, \leq)$ where

$$\mathcal{D} = \{ H \mid H \subseteq \mathcal{H} \}$$

and $D_1 \leq D_2$ iff $D_1 \subseteq D_2 \; \forall \; D_1, D_2 \; \varepsilon \; \mathcal{D}$.

The semantics of statecharts is given by a semantic function M that maps any statechart which is a member of U, the set of all statecharts, to an element of D, ie, M: U $\rightarrow$ D. The semantics is an a priori semantics that anticipates an arbitrary environment.

3. FNLOG: OVERVIEW

The main building blocks of a specification in FNLOG are events and activities. An event is an instantaneous occurrence of a signal. An activity is a durative happening, with a beginning instant, an end instant and a finite duration between the two instants. For example, *move* is an activity whereas *stop* is an event for a moving object. Events and activities may arise within the system or in the external environment. Every event e and activity A is superscripted by a number i, written as e^i and A^i, which indicates that it is the ith occurrence of the event or activity in the current system incarnation. This notation permits the specification of repeated events and activities over time.

With every activity A, we associate two special events, called *init-A* and *term-A*, corresponding to the initiation and termination of activity A. Primitive events and activities are abstractions which are left unspecified. Pure signals are primitive events.

The logical operators of $\wedge$, $\vee$, $\neg$, $\rightarrow$, $\leftrightarrow$ are included in our language to facilitate composition of events and activities into higher level events and activities. For convenience, we use only the first three; the other two are derivable from them. We also use temporal descriptors to capture relative time properties of events and activities. These descriptors are the past time temporal logic operators:

(i) $\odot_t$ true at time t
(ii) $\ominus_t$ true at the instant previous to t
(iii) $\Diamond_t$ true at some instant before t
(iv) $\Box_t$ true at all instants before t

The temporal operators are applicable to both events and activities. The usage of t is that of a variable which takes any permissible value. As implied already, our concept of time is that of an infinite sequence of discrete time instants.

We employ hierarchical composition of events and activities to derive the higher level events and activities. Higher level events and activities, which are of greater complexity than the primitive ones, are composed of logical and temporal predicates which directly or derivatively employ the primitive events and activities. Thus a hierarchy of events and activities may be built.

The existential and universal quantifiers are allowed to range over the time variable t. Actually, even the temporal operators are short-hand notations to indicate range over time t. We borrow from quantified temporal logic (Pnueli and Manna, 1988) and introduce quantified temporal operators as short-hand for quantification:

(i) $\odot_{t-k}$ true k instants before time t
(ii) $\Diamond_t^{t-k}$ true at some instant in the interval [t-k, t]
(iii) $\Box_t^{t-k}$ true at all instants in the interval [t-k, t]

3.1 *Syntax of FNLOG*

For defining a formal syntax of FNLOG, we define the following sets:

E_e = set of elementary/atomic events generated by system and environment

E = set of all events generated by system and environment

$E_e \subseteq E$

A_e = set of elementary atomic activities performed by system and environment

A = set of all activities generated by system and environment

$A_e \subseteq A$

For every a ε A, init-a, term-a ε E

Let a, b, a_1,... ε A, e, e_1,.... ε E

Reserved Variables. The language contains special variables called reserved variables to correspond to the observables and non-observables of the system. The observables are the events generated by the system at each time instant, while the non-observables denote the set of events generated by the whole system including the environment at each step and the causality relationship between generated events. Our language contains reserved variables corresponding to each of these denotations:

(i) t_b: corresponds to the system entry instant
(ii) t_e: corresponds to the system exit instant
(iii) τ_b: corresponds to the system entry event
(iv) τ_e: corresponds to the system exit event
(v) S: corresponds to the set of all events generated by the system at a given instant
(vi) $<<$: corresponds to the causality relation between events occurring simultaneously at a given instant.

Logical Variables. The logical variable environment γ assigns values to free logical variables occurring in the specifications and assertions of the language. There are five types of logical variables: the N-variables denote time, the T-variables denote transitions, the F-variables denote an event or activity, the B-variables denote the truth or falsity of the formulae and the E-variables designate the set of events occurring at an instant.

(i) logical N-variables which range over the set N_0 = N $\cup$ {0}. Symbols are t, t_1, n, m,.... Quantification over the N-variables is allowed. Let LNvar be the set of logical N-variables.
(ii) logical transition variables which range over the set $\tau \cup$ {*, $\perp$ }. Symbols are T, T_1, ... Let LTvar be the set of logical T-variables.
(iii) logical function variables, denoting an event or activity, called collectively as a function. Symbols are f, f_1,.... Let LFvar be the set of logical function variables.

(iv) logical boolean variables, denoting a value in {True, False}. Let LBvar be the set of logical boolean variables.
(v) logical event set variables, ranging over functions from N to sets of events. Symbols are s, s_1, ... Let LEvar be the set of logical event set variables.

Operators. The operators are of two types: logical and temporal.

(i) The logical operators are $\wedge$, $\vee$, $\neg$.
(ii) The temporal operators, denoted in general by T_t, are $\odot_t$, $\ominus_t$, $\Diamond_t$, $\square_t$, $\odot_{t-k}$, $\Diamond_t^{t-k}$, $\square_t^{t-k}$.

The Specification. To specify any system using FNLOG, all the activities a which form part of the system are identified. The top level specification is defined as an OR or AND of these activities. Each activity is then refined in a top down manner using other simpler activities and events as building blocks. This is a multi-staged process, with the simpler activities and events themselves capable of being likewise refined. Thus, a sequence of definitions of the activities over time is given, by utilizing the temporal and logical operators to connect them. The syntax of a specification is dealt with next.

The Syntax. We define five types of expressions corresponding to the five types of logical variables:

1. Number expressions denote a value in N_0. Their syntax is given by N-variables connected by arithmetic operators +. -, *. Let t ε LNvar.

exp ::= 0 | 1 | t | exp_1 + exp_2 | exp_1 - exp_2 | exp_1 * exp_2

2. Transition expressions denote a value in $\tau \cup$ {*, $\perp$ }. Let TT ε LTvar, T ε τ.

texp ::= * | $\perp$ | T | TT

3. Function expressions denote a function in A $\cup$ E.

For f ε LFvar, e ε E, a ε A,
fexp ::= e | a | init-a | term-a | f

4. Boolean expressions denote a value in { True, False }. The operators of $\vee$, $\wedge$, $\neg$ are used to form the boolean expressions.

bexp ::= True | False | exp_1 = exp_2 | exp_1 < exp_2 | exp_1 > exp_2 | $\neg$ bexp | $bexp_1$ $\vee$ $bexp_2$ | $bexp_1$ $\wedge$ $bexp_2$ | e_1 $<<_{exp}$ e_2

5. Event set expressions denote subsets of events. Let A $\subseteq$ E, s ε LEvar.

eset ::= A | S(exp) | s(exp) | $eset_1 \subseteq eset_2$ | $eset_1$ $\cup$ $eset_2$ | $eset_1$ - $eset_2$

Let L be the logic-based functional specification of the system. Then

L::= $\vee_a \odot_t a$ | $\wedge^a \odot_t a$

where the a's are defined by temporal formulae. In general, let f ε A $\cup$ E be any function. Then

$\odot_t f ::= tformula$

where the temporal formula *tformula* is defined by

tformula ::= True | False | T_t (bexp) | T_t (e) | T_t (a) | ¬ tformula | $tformula_1 \wedge tformula_2$ | $tformula_1 \vee tformula_2$ | e ε eset | $eset_1$ = $eset_2$ | bexp | exp: tformula (exp) | $\exists$ exp: tformula (exp)

Interpretation of γ. The logical variable environment γ assigns values to the free variables in the tformulae, ie to variables in the expressions.

γ: (exp $\rightarrow$ N) x (texp $\rightarrow$ τ $\cup$ {*, $\perp$}) x (fexp $\rightarrow$ E $\cup$ A) x (bexp $\rightarrow$ {T, F}) x (eset $\rightarrow$ $\mathcal{P}$ (E)).

4. SEMANTICS AND PROOF SYSTEM

Our aim is to specify a compositional proof system for statecharts. As a first step, we define FNLOG specifications corresponding to every statecharts component, such that their semantics is preserved. This is achieved by defining for FNLOG a semantics that maps primitive events and activities onto subsets of histories of a corresponding statechart. By this means and by exploiting the compositionality of statecharts from syntactic components, we are able to generate for every statechart, a corresponding FNLOG specification.

In this section, we briefly describe the technique for the basic statechart and indicate the procedure for a few other components.

4.1 *Semantics of FNLOG*

We define a compositional semantics for FNLOG based on the semantics of statecharts. Let F and U be the FNLOG and statecharts specifications of any system. We associate subsets of history tuples from the domain of the statecharts semantic function, with events and activities of FNLOG operated upon by logical and temporal operators. Some details are omitted for simplicity.

The interpretation of a formula in FNLOG with respect to a statechart maps the reserved variables of FNLOG to those of statecharts and the logical variables to their evaluations in the logical environment.

An event e occurring at time n is mapped to the subset of all histories in which e occurs at the nth step. Thus,

$[[\odot_n (e)]] = \{$ h ε $\mathcal{H}|$ e ε f^C (n) $\}$

For a time variable t, we must consider all possible values of t and collect the corresponding histories in which e occurs. Thus

$[[\odot_t (e)]] = \bigcup_{n \varepsilon N} \{$ h ε $\mathcal{H}$ | e ε f^C (n) $\}$
$= \bigcup_{n \varepsilon N} [[\odot_n (e)]]$

An activity a is defined by its initiating event *init-a*, terminating event *term-a* and the durative component. At any instant, an activity a is occurring if it was previously initiated and has not yet terminated. Thus we have

$[[\odot_n (\text{init-a})]] = \{$ h ε $\mathcal{H}$ | init-a ε f^C (n) $\}$

$[[\odot_n (\text{term-a})]] = \{$ h ε $\mathcal{H}$ | term-a ε f^C (n) $\}$

$[[\odot_n (a)]] = \{$ h ε H | $\exists$ $n_1 \leq$ n : init-a ε f^C (n_1) $\wedge$ $\forall$ n_2, $n_1 \leq n_2 \leq$ n : ¬ (term-a ε f^C (n_2)) $\}$

As before, for variable t we have

$[[\odot_t (a)]] = \bigcup_{n \varepsilon N} [[\odot_n (a)]]$

The semantics of the temporal operators applied on events and activities may be easily derived, as may the semantics of the logical operations. Briefly, $\diamond_n$ maps to union of history subsets, while $\square_n$ maps to intersection of history subsets. The logical $\wedge$ and $\vee$ map to intersection and union of history subsets respectively, while ¬ maps to the complement.

4.2 *Mappings*

First we consider the basic statechart B = [I, O, A]. Entry to B is either implicit (denoted by *) or via one of the arcs in I at a particular time step. From the next step onwards, three situations are possible:

1. wait inside state A forever (denoted by $\perp$ = 'never exit')
2. exit at a finite step implicitly (also denoted by *)
3. take one of the outgoing arcs in O, for which the necessary condition for taking the transition should be true.

The FNLOG specification of B must capture these ideas. Let L_B be the FNLOG specification of B. Let B = [I,O,A] where I $\subseteq$ τ_I, O $\subseteq$ τ_O, A ε $\sum$

For T_i ε I $\cup$ { * }, define

$\odot_n (T_i) = (\tau_b = T_i) \wedge (t_b = n)$

Define $\odot_n$ (init-A) $= \bigvee_{T_i \varepsilon I \cup \{*\}} \odot_n (T_i)$

For $T_k \varepsilon O$, let $T_k = (A, l_k, -)$ where

A = source state of T_k

$l_k = E_k/A_k$ is the label of T_k; we omit the condition part for simplicity.

For $T_k \varepsilon O \cup \{ *, \bot \}$, define

$\odot_n (T_k) = (t_e = n) \wedge \Box_{t_b}^n (\odot_v (\text{wait}(O))) \wedge \odot_n (\text{fire}(T_k))$

$\odot_v (\text{wait}(O))$ characterizes the situation in which no transition in O can be taken; then no trigger evaluates to true and no event is generated. $\odot_v$ (fire(O)) describes the condition for taking a transition T_k in O at step v.

Define $\odot_n$ (term-A) $= \bigvee_{T_k \varepsilon O \cup \{*, \bot\}} \odot_n T_k)$

Finally, define $L_B = \odot_n$ (A) where

$\odot_n$ (A) $= \odot_{n_1}$ (init-A) $\wedge \odot_{n_2}$ (term-A) $\wedge$ (n1 $\leq$ n $\leq$ n_2)

Mapping for Or. Let L_i be the logical specification of U_i. By definition of Or (U_1, U_2),

$\tau_I^{U_1} \cap \tau_I^{U_2} = \Phi$

$\tau_O^{U_1} \cap \tau_O^{U_2} = \Phi$

Thus the logical specification of Or (U_1, U_2) is given by

1. $L = L_1 \vee L_2$
2. All formulas of L_1 and L_2 hold.

Mapping for And. By definition of And (U_1, U_2),

$\tau_I^{U_1} \cap \tau_I^{U_2} = \Phi$

$\tau_O^{U_1} \cap \tau_O^{U_2} = \Phi$

Thus the logical specification of And (U_1, U_2) is

$L = L_1 [\tau_{b_1} / \tau_b, S_1 / S, \tau_{e_1}/ \tau_e] \wedge [\tau_{b_2} / \tau_b, S_2 / S, \tau_{e_2}/ \tau_e] \wedge \odot_n$ (and-in) $\wedge \odot_n$ (and-out) $\wedge$ S $= S_1 \cup S_2$

with τ_{b_1}, τ_{b_2}, τ_{e_1}, τ_{e_2}, S_1 and S_2 fresh variables and

$and\text{-}in = (\tau_b = \tau_{b_1} \wedge \tau_{b_2} = *) \vee (\tau_b = \tau_{b_2} \wedge \tau_{b_1} = *)$

$and\text{-}out = (\tau_e = \tau_{e_1} \neq \bot \wedge (\tau_{e_2} = *) \vee \tau_e = \tau_{e_2} \neq \bot \wedge \tau_{e_1} = *) \vee (\tau_e = \tau_{e_1} = \tau_{e_2} = \bot)$

Mapping for Stat. Consider Stat (B, U, T) which makes B a superstate with U inside it, and $T \varepsilon \tau_I^U$ as its default. Any entry to B leads to U via the default arc T and direct entry via T is no longer possible. Let the logical specifications of B and U be L_B and L_U . Then the FNLOG specification of Stat (B, U, T) is given by

$L = L_B [\tau_{b_1} / \tau_b, S_1/S, \tau_{e_1}/ \tau_e] \wedge L_U [\tau_{b_2}/ \tau_b, S_2/S, \tau_{e_2}/ \tau_e] \wedge \odot_n$ (stat-in) $\wedge \odot_n$ (stat-out) $\wedge$ S $= S_1 \cup S_2$

with $\tau_{b_1}, \tau_{b_2}, \tau_{e_1}, \tau_{e_2}, S_1$ and S_2 fresh variables and

$stat\text{-}in = (\tau_e = \tau_{e_1} \wedge \tau_{e_2} = T) \vee \tau_e \neq * \wedge \tau_e \neq T \wedge \tau_e = \tau_{e_2} \wedge \tau_{e_1} = *)$

$stat\text{-}out = stat\text{-}in$

The mappings for Connect and HiCl are more complicated and (Sowmya and Ramesh, 1994) may be consulted for details.

4.3 Semantic Equivalence

This is now easy to show, based on the mappings and the semantics of events and activites.

For the basic statechart B = [I,O,A], the interpretation of reserved variables with respect to a history and $\mathcal{M}$, the semantic function of statecharts mapping to sets of histories, we have

$[[\odot_n$ (A)$]] = [[\odot_{n_1}$ (init-A)$]] \cap [[\odot_{n2}$ (term-A)$]] \cap [[n_1 \leq n \leq n_2]]$

$= \{ h \varepsilon \mathcal{H} \mid s = n_1 \wedge i \varepsilon I \cup \{*\}\} \cap \{h \varepsilon \mathcal{H} \mid s = n_2 \wedge \forall v, s < v < s: \odot_v$ (wait(O)) $\wedge [o = \bot \vee (o = * \wedge f^F (s) = \phi) \vee \odot_s$ (fire(O))$] \cap \{h \varepsilon \mathcal{H} \mid s \leq s \}$

We can show that

$[[\odot_v$ (wait(O))$]] = \{h \varepsilon \mathcal{H} \mid$ wait(O, v) $\}$

and $[[\odot_v$ (fire(O))$]] = \{h \varepsilon \mathcal{H} \mid$ fire(O, v) $\}$

Thus $[[\odot_n$ (A)$]] = \{h \varepsilon \mathcal{H} \mid i \varepsilon I \cup \{*\} \wedge \forall v, \hat{s} < v < s:$ wait (O, v) $\wedge [o = \bot \vee (o = * \wedge f^F (s) = \phi) \vee$ (fire(O, s))$]\}$

$= \mathcal{M}$ ([I, O, A])

Equivalence for Or. Since L_i is the FNLOG specification of U_i generated by our mappings, $[[L_i]] = \mathcal{M}$ (U_i). Thus, for the FNLOG specification of Or(U_1, U_2) defined earlier,

$[[L]] = [[L_1]] \cup [[L_2]] = \mathcal{M}$ (U_1) $\cup \mathcal{M}$ (U_2)

Equivalence for And. For And (U_1, U_2), and its FNLOG specification L defined earlier, we have

$[[L]] = [[L_1 [\tau_{b_1} / \tau_b, S_1/S, \tau_{e_1}/ \tau_e]]] \cap [[L_2 [\tau_{b_2}/ \tau_b, S_2/S, \tau_{e_2}/ \tau_e]]] \cap [[\odot_n$ (and-in) $]] \wedge [[\odot_n$ (and-out) $\wedge [[S = S_1 \cup S_2]]$

$[[L_1 [\tau_{b_1}/ \tau_b, S_1/S, \tau_{e_1}/ \tau_e]]] \cap [[L_2 [\tau_{b_2}/ \tau_b, S_2/S, \tau_{e_2}/ \tau_e]]]$

$= \{ h \varepsilon \mathcal{H} \mid \exists h_1 \varepsilon [[L_1]], h_2 \varepsilon [[L_2]]: \hat{s} = \hat{s}_1 = \hat{s}_2 \wedge s = s_1 = s_2 \wedge f^C = f_1^C = f_2^C \wedge f^< = f_1^< = f_2^< \}$

$[[S = S_1 \cup S_2]] = \{ h \varepsilon \mathcal{H} \mid \exists h_1 \varepsilon [[L_1]], h_2 \varepsilon [[L_2]]: f^F = f_1^F \cup f_2^F \}$

$[[\odot_n \ (and\text{-}in)]] = \{ \ h \ \varepsilon \ \mathcal{H} \ | \ \exists \ h_1 \ \varepsilon \ [[L_1]], \ h_2 \ \varepsilon$
$[[L_2]]: (i = i_1 \wedge i_2 = *) \vee (i = i_2 \wedge i_1 = *)\}$

$[[\odot_n \ (and\text{-}out) = \{ \ h \ \varepsilon \ \mathcal{H} \ | \ \exists \ h_1 \ \varepsilon \ [[L_1]], \ h_2 \ \varepsilon \ [[L_2]]:$
$(o = o_1 \neq \perp \wedge o_2 = *) \vee (o = o_2 \neq \perp \wedge o_1 = *)$
$\vee \ (o = o_1 = o_2 = \perp) \ \}$

Hence $[[L]] = \mathcal{M} \ (And \ (U_1, \ U_2))$

<u>Equivalence for Stat.</u> This follows exactly similar to And except for the redefinitions of *stat-in* and *stat-out*.

5. PROOF SYSTEM

To verify a statecharts specification, we first transform it to an equivalent FNLOG specification using the mappings defined above and then verify the FNLOG specification. Our proof system for FNLOG is deductive and logic-based.

5.1 *Axioms*

The axioms consist of the axioms of first-order predicate logic, axioms of past time temporal logic and the axioms relating function, event and activity definitions in the application domain. Of these, we discuss the last category here; the others are standard.

For all function definitions of the form $\odot_n \ (f)$ = tformula, we have the *definitional axiom* as follows:

$[\odot_t \ (f) = \text{tformula}] \longleftrightarrow [\odot_t \ (f) \rightarrow \text{tformula} \wedge$
$\text{tformula} \rightarrow \odot_t \ (f)]$

Since the basic building blocks of FNLOG, the *events* and *activities*, are also related causally and temporally, we may define an axiomatic system encoding these relationships. It arises from the fact that each activity comprises an initiating event and a terminating event, besides a durative component. We shall call these axioms the *application level axioms*. They may be derived directly from the axioms of arithmetic, given the causal relationships. For example, if an activity a is true at a given instant, then it must have been instantiated some time in the past. Thus,

(a) $\odot_t \ (a^i) \rightarrow \Diamond_t \ (\text{init-}a^i)$

The other axioms follow:

(b) $\odot_t \ (a^i) \rightarrow \neg \ \Diamond_t \ (\text{term-}a^i)$
(c) $\odot_t \ (\text{init-}a^i) \rightarrow \neg \ \odot_t \ (\text{term-}a^i)$
(d) $\odot_t \ (\text{init-}a^i) \rightarrow \odot_t \ (a^i)$
(e) $\neg \ \Diamond_t \ (\text{init-}a^i) \rightarrow \neg \ \odot_t \ (a^i)$
(f) $\neg \ \odot_t \ (\text{term-} a^i) \wedge \Diamond_t \ (\text{init-}a^i) \ arrow \ \odot_t \ (a^i)$
(g) $\neg \ \odot_t \ (a^i) \rightarrow \neg \ \odot_t \ (\text{init-}a^i)$
(h) $\odot_t \ (\text{term-}a^i) \rightarrow \odot_t \ (a^i)$

(i) $\odot_t \ (\text{term-}a^i) \rightarrow \Diamond_t \ (\text{init-}a^i)$
(j) $\neg \ \Diamond_t \ (\text{init-}a^i) \rightarrow \neg \ \odot_t \ (a^i)$ (k) $\neg \ \Diamond_t \ (\text{init-}a^i)$
$\rightarrow \neg \ \Diamond_t \ (a^i \ )$
(l) $\neg \ \Diamond_t \ (a^i) \rightarrow \neg \ \odot_t \ (\text{init-}a^i \ )$

Further, since multiple occurrences of the same event/activity are permitted, some axioms also capture the consequent relationships; we shall call these the *repeated occurrence axioms*. For example, the following axiom states that any event or activity can occur repeatedly but sequentially. For example,

$\forall \ i, \ j, \ \odot_t \ (a^i) \wedge \odot_{t'} \ (a^j) \wedge i < j \rightarrow t < t'$

The basic repeated occurrence axioms are:
(a) $\odot_t \ (a^i) \rightarrow t \geq 0, i \geq 1$
(b) $\forall \ i, \ j \ \odot_t \ (a^i) \wedge \odot_{t'} \ (a^j) \wedge i < j \rightarrow t < t'$

The axioms that follow capture the basic property of repeated occurrences of a single event or activity, namely that the $i+1^{th}$ occurrence strictly follows the ith occurrence in time:
(c) $\odot_t \ (a^i) \rightarrow \neg \ \Diamond_t \ (\text{init-}a^{i+1})$
(d) $\odot_t \ (a^i) \rightarrow \neg \ \Diamond_t \ (a^{i+1})$
(e) $\odot_t \ (a^i) \rightarrow \neg \ \Diamond_t \ (\text{term-}a^{i+1})$
(f) $\odot_t \ (\text{init-}a^i) \rightarrow \neg \ \Diamond_t \ (\text{init -}a^{i+1})$
(g) $\odot_t \ (\text{init-}a^i) \rightarrow \neg \ \Diamond_t \ (a^{i+1})$
(h) $\odot_t \ (\text{init-}a^i) \rightarrow \neg \ \Diamond_t \ (\text{term -}a^{i+1})$
(i) $\odot_t \ (\text{term-}a^i) \rightarrow \neg \ \Diamond_t \ (\text{init-}a^{i+1})$ (j) $\odot_t \ (\text{term-}a^i) \rightarrow \neg \ \Diamond_t \ (a^{i+1})$
(k) $\odot_t \ (\text{term-}a^i) \rightarrow \neg \ \Diamond_t \ (\text{term-}a^{i+1})$
(l) $\odot_t \ (\text{init-}a^{i+1}) \rightarrow \Diamond_t \ (\text{init-}a^i)$
(m) $\odot_t \ (\text{init-}a^{i+1}) \rightarrow \Diamond_t \ (a^i)$
(n) $\odot_t \ (\text{init-}a^{i+1}) \rightarrow \Diamond_t \ (\text{term-}a^i)$
(o) $\odot_t \ (\text{term-}a^{i+1}) \rightarrow \Diamond_t \ (\text{init-}a^i)$
(p) $\odot_t \ (\text{term-}a^{i+1}) \rightarrow \Diamond_t \ (a^i)$
(q) $\odot_t \ (\text{term-}a^{i+1}) \rightarrow \Diamond_t \ (\text{term-}a^i)$
(r) $\odot_t \ (a^{i+1}) \rightarrow \Diamond_t \ (\text{init-}a^i)$
(s) $\odot_t \ (a^{i+1}) \rightarrow \Diamond_t \ (a^i)$
(t) $\odot_t \ (a^{i+1}) \rightarrow \Diamond_t \ (\text{term-}a^i)$

5.2 *Proof Rules*

All proof rules of first-order predicate logic and past-time temporal logic hold and are not repeated here. In addition, we have new application level proof rules which are derived from the semantics of events and activities described in the previous section.

From the semantics of the temporal operations on events defined earlier, we obtain the following rules, which are in the conventional form, as may be expected. We present the rules for single occurrence of events and activities; extensions to multiple occurrences are straightforward.

Let $e, \ e_1,....$be events, $T, \ T_1 \ ,....$temporal operators, $a, \ a_1 \ ,....$activities and $\phi, \ \phi_1,....$FNLOG assertions. Then we have the rules corresponding to the temporal operations on events:

1. $\odot_n (e) \rightarrow \odot_n(e)$

2. $\odot_{n-1} (e) \rightarrow \phi$

$\ominus_n (e) \rightarrow \phi$

3. $\exists\, i,\, t_b \leq i \leq n:\, \odot_i (e) \rightarrow \phi_i$

$\diamondsuit_t (e) \rightarrow \bigvee_{i=t_b}^{n} \phi_i$

4. $\forall i, t_b \leq i \leq n:\, \odot_i (e) \rightarrow \phi_i$

$\square_t (e) \rightarrow \bigwedge_{i=t_b}^{n} \phi_i$

For the logical operations on events, we have defined the semantics by means of set operations, which translate to logical operations in the proof rules. Set union, intersection and complement map to logical Or, And and not. Thus the corresponding proof rules are:

5. $T(e_1) \rightarrow \phi_1,\ T(e_2) \rightarrow \phi_2$

$T (e_1 \wedge e_2) \rightarrow \phi_1 \wedge \phi_2$

6. $T(e_1) \rightarrow \phi_1,\ T(e_2) \rightarrow \phi_2$

$T (e_1 \vee e_2) \rightarrow \phi_1 \vee \phi_2$

7. $\odot_n (e) \rightarrow \phi$

$\odot_n (\neg e) \rightarrow \neg \phi$

8. $\ominus_n (e) \rightarrow \phi$

$\ominus_n (\neg e) \rightarrow \neg \phi$

9. $\square_n (e) \rightarrow \phi,\ \phi = \wedge_i \phi_i,\ \odot_i (e) \rightarrow \phi_i$

$\diamondsuit_n (\neg e) \rightarrow \neg \phi \rightarrow \vee_i \neg \phi_i$

10. $\diamondsuit_n (e) \rightarrow \phi,\ \phi = \vee_i \phi_i,\ \odot_i (e) \rightarrow \phi_i$

$\square_n (\neg e) \rightarrow \wedge_i \neg \phi_i$

Also, for logical operations on top of temporal operations, we have similar rules:

11. $T_i (e_i) \rightarrow \phi_i,\ i = 1, 2$

$T_1 (e_1) \wedge Te_2) \rightarrow \phi_1 \wedge \phi_2$

12. $T_i (e_i) \rightarrow \phi_i,\ i = 1, 2$

$T_1 (e_1) \vee T(e_2) \rightarrow \phi_1 \vee \phi_2$

13. $\odot_n (e) \rightarrow \phi$

$\neg \odot_n (e) \rightarrow \neg \phi$

14. $_n (e) \rightarrow \phi$

$\neg \ominus_n (e) \rightarrow \neg \phi$

15. $\diamondsuit_t (e) \rightarrow \phi,\ \phi = \vee_i \phi_i,\ \odot_i (e) \rightarrow \phi_i$

$\neg \diamondsuit_t (e) \rightarrow \neg (\vee_i \phi_i) \rightarrow \wedge_i \neg \phi_i$

16. $\square_t (e) \rightarrow \phi,\ \phi = \wedge_i \phi_i,\ \odot_i (e) \rightarrow \phi_i$

$\neg \square_t (e) \rightarrow \neg (\wedge_i \phi_i) \rightarrow \vee_i \neg \phi_i$

In addition, we have similar proof rules based on the semantics of an activity. To avoid tedium, they are not repeated here. Note, however, that in the case of activities, it is necessary to know the instant when the activity was last initiated. This may be extracted from the semantic histories. The rules for activities incorporate this fact.

6. CONCLUSION

In this paper, we have described a semantics-preserving transformation of statecharts to FN-LOG and a proof system for FNLOG based on the semantics. It is pointed out that the semantics turns out to be intuitive, in that logical disjunctions and conjunctions do map to set union and intersection in the semantic domain. The proof system builds on first order predicate and first order temporal logic, and hence is able to reuse their axioms. The additional axioms arise due to the introduction of the primitive events and activities into the language.

7. REFERENCES

Harel, D. (1987). Statecharts: a visual approach to complex systems. *Science of Computer Programming* **8**(3), 231–274.

Hooman, J., S. Ramesh and W. P. de Roever (1989). A compositional semantics for statecharts. In: *Proc. Formal models of concurrency, Novosibirsk, USSR.*

Huizing, C., R. Gerth and W. P. de Roever (1988). Modelling statecharts behaviour in a fully abstract way. In: *CAPP88 Lecture Notes in Computer Science* (M. Dauchet and M. Nivat, Eds.). Vol. 299. Springer Verlag. Berlin.

Pnueli, A. and Z. Manna (1988). Applications of temporal logic to the specification of real-time systems. In: *Formal Techniques in real-time and fault-tolerant systems* (M. Joseph, Ed.). Vol. 331. pp. 84–98. Springer Verlag. Berlin.

Sowmya, A. and S. Ramesh (1992). Verification of timing properties in a statecharts- based model of real-time reactive systems. In: *Distributed Computer Control Systems, IFAC Workshop Series 1992* (H. Kopetz and M. G. Rodd, Eds.). Vol. 3. Pergamon Press. Oxford.

Sowmya, A. and S. Ramesh (1994). Extending statecharts with temporal logic. Technical report. School of Comp. Sci and Engg, University of New South Wales, Sydney.

A FAULT-TOLERANT ARCHITECTURE FOR LARGE-SCALE DISTRIBUTED CONTROL SYSTEMS

Holger Hilmer* Hans-Dieter Kochs* Ewald Dittmar**

*Department of Computer Science, University of
Duisburg/Germany
** ABB Netzleittechnik GmbH, Network Control and Protection,
Ladenburg/Germany

Abstract: This article introduces a DCCS architecture and protocol mechanisms designed for application areas for which the major attributes are high reliability and hard real-time operation. The application domains addressed are characterized by a large number of network nodes and a considerable amount of sporadically occurring data. Design rules are derived from the fundamentals of distributed real-time computing and the requirements of the target application domains leading to an advantageous system concept. This concept provides an active error detection and notification mechanism, as opposed to the fail-silent paradigm. The system architecture makes use of existing standards and components where possible.

Keywords: Real-Time Communication, Fault-Tolerant Systems, Distributed Computer Control Systems, Communication Protocols, Reliability, Recovery Times

1. INTRODUCTION

Concerning communication, distributed control systems at the lowest level of the process control hierarchy do have important common characteristics: mostly cyclic data with short information lengths have to be transfered under hard real-time constraints. In addition, many applications are required to ensure dependable system operation by means of fault tolerance techniques. Highly reliable communication protocols and architectures (e.g. MARS (Kopetz, 1995)) often make use of periodic bus access strategies, such as TTP, token, or polling methods which from the performance viewpoint are very efficient transferring periodically occurring data. But even low level control systems have to cope with sporadic data triggered by uncontrolled external events, such as process state changes, process alarms, and component faults. Periodic communication protocols have to reserve bandwidth of the bus medium in order to transmit those spontaneous messages with mini-

mum delay. In case the amount of sporadic data exceeds a specific limit, periodic access strategies are not appropriate since the average access delays increase due to the required amount of reserved bandwidth. As a result, real-time operation cannot be guaranteed. Especially regarding large-scale networks, bandwidth is limited because a lot of nodes have to be regarded in the pre-calculated schedule. Here, an event-triggered multi-master protocol (CSMA protocols: ETHERNET, CAN, LON) seems to be the best choice. But, existing multi-master protocols do not provide both real-time behavior and high reliability. On one hand, highly reliable architectures are optimized for use in high level control centers in which large amounts of data have to be transfered under low real-time constraints. Here, often protocols based on ETHERNET are used, not appropriate to transfer small messages with high frequency. On the other hand, protocols sufficient for hard real-time operation, e.g. fieldbus protocols, require additional measures concerning fault tolerance.

Fault tolerance comprises hardware redundancy and demands measures for error processing such as error detection, localization, notification, and redundancy management. A main goal of a fault-tolerant distributed system is that redundancy switch-over takes place without loss, corruption, and duplication of messages. Moreover, error processing may not effect real-time operation seriously. Periodic protocols and multi-master systems require different fault tolerance techniques, especially regarding error detection and notification. For example, using a periodic protocol a node loss is detected automatically since the pre-calculated sending schedule will be disturbed. Existing multi-master protocols provide mechanisms in which a node loss is detected through the use of so-called live-guarding methods. Life guarding is used to refer to the cyclical transmission of life messages by all of the operational nodes to a master node. If a life message does not occur, that indicates a node loss. Depending on the cycle time of the life messages, an unacceptably long time may pass until the node failure is detected so that a loss of messages occurs leading to inconsistencies.

Regarding event-triggered multi-master protocols, due to high error detection and notification latencies, the well-known fail-silent strategy is not sufficient. Therefore, an active error mechanism has to be provided for detecting and indicating a component fault with a minimal delay. In the following, a DCCS architecture is described including fault tolerance services according to the above design rules.

2. FUNDAMENTALS OF DISTRIBUTED SYSTEM DESIGN

2.1 Real-Time Communication

Real-time communication in a distributed system demands every message to be transmitted within its allocated worst case transaction time. The transaction time comprises the delays from the occurrence of an information to the reception of the message at the receiving nodes. Thus, real-time communication requires deterministic data transfer, short processing times and an adequate transmission frequency (bandwidth). The performance requirements are strongly determined by the most time-critical informations, which have to be transmitted with a very low latency jitter, i.e. the maximum allowed deviation of the assigned transaction time is very small.

The bus access strategy of the communication protocol exerts a great influence on the timing behavior of data exchange. The access method should be designed according to the characteristic of data and the real-time requirements. The data characteristic means the way events occur (sporadically or periodically), the length of informations, the number of sending and receiving nodes and the timing constraints communication is subjected to. There are two major classes of bus access principles protocols can be assigned to: on one hand event-driven methods such as CSMA (Carrier Sense Multiple Access) and its variants, on the other hand time-driven protocols comprising token, polling and time-slice methods. A network node operating an event-driven protocol is allowed to start the transmission of a message at any point of time, unless the bus is occupied by another transmission activated before. Time-driven protocols are characterized by an a priori known time schedule, i.e. the time and the sequence of bus access is assigned to all network nodes off-line.

Each of the mechanisms offers advantages and disadvantages with respect to different real-time and reliability requirements as well as different data characteristics. Event-driven protocols provide very low bus access delays on the average, unless the busload exceeds a maximum value. However, determinism of bus access is limited, because in case several nodes try to occupy the bus at the same time the access delays cannot be determined exactly. Some event-oriented protocols improve the determinism capability through the use of priority controlled bus access methods.

Time-driven protocols show worse average access times, especially regarding networks with a high number of nodes, but latency jitter is small, since the access delays of all nodes are fixed during system design.

Using time-oriented bus access methods different problems occur, which reduce the effective bandwidth of the bus. In case sporadic messages triggered by unforeseen events have to be transfered, which cannot be considered by the off-line time schedule, bandwidth has to be reserved. That means, the sending time assigned to every node within a transmission cycle in order to transmit the periodic informations has to be enlarged by a sufficient amount of time to send possible sporadic messages. If no event occurs during the sending time of a network node, bandwidth has been wasted. This affects the average bus access delays negatively, especially regarding systems with a high number of nodes and a considerable amount of sporadic data to be transfered.

Concerning real-time behavior, an important aspect is the protocol efficiency. It means the ratio of the bandwidth used for the transmission of process data to the overall busload caused by the protocol mechanisms. The protocol efficiency E_p is defined as:

$$E_p = \frac{\sum_1^n L_{pi}}{\sum_1^n L_f + \sum L_{pc}} * 100\% \qquad (1)$$

Where n is the number of messages transfered during a sufficient period of time. L_{pi} describes the length of a process information. L_f is the total frame length including process informations as well as frame overhead such as address informations, checksum bits, and other protocol specific informations. The most significant difference between protocol implementations is the overhead caused by protocol control messages L_{pc}. For example, this includes token and polling frames in order to control the bus access as well as acknowledgment messages confirming the correct reception of messages. In general, event-driven protocols, as opposed to time-driven systems, provide a lower control overhead, since the bus access demands no additional use of bandwidth.

Concerning real-time operation the above aspects lead to the following requirements a protocol has to meet:

- event-driven bus access with priority control
- high protocol efficiency
- sufficient bandwidth of the communication system

2.2 Fault-Tolerant Communication

The second basic technical requirement communication has to meet is high reliability, which has gone unnoticed so far in many applications. Highly reliable system operation can be realized by fault-tolerant system design including hardware redundancy and mechanisms to manage error handling. Fault tolerance means that the functioning of the entire system is to be maintained despite faulty components.

Distributed systems deal with the multiplication and distribution of information to locally separated function modules. Multiplication and distribution must take place consistently, i.e. an item of source information must be present at the receivers in an identical state within a specific time. Inconsistencies of the distributed databases caused by faulty components can lead to a severe malfunction of the entire system. In conjunction with consistency, fault tolerance signifies that even when a fault occurs data consistency is to be preserved or restored before a propagation of faults affects the overall system function. Especially in systems dealing with data mostly occurring spontaneously, i.e. event-triggered, measures have to be taken to ensure data consistency. Due to the fact that generally the states of all

databases are regenerated within a transmission cycle, periodic data characteristics cause less difficulties since after a component fault system consistency is recovered automatically.

Inconsistencies can be avoided or at least detected through the use of a sufficient transmission strategy. Moreover, event-oriented systems, as opposed to periodic protocols, require an acknowledgment mechanism in order to detect a message loss. With respect to the acknowledgment process, the degree of reliability increases with the number of confirming receivers. The maximum possible reliability of the transmission of data is obtained with the atomic broadcast principle: a message is either correctly received by all of the operationally capable network nodes or it is received by none of them. This principle can be realized by multiphase acknowledgment methods, comprising a sequential exchange of confirmation messages between the sender and all of the receivers until finally, the sender enables the message processing at the receiving nodes. These concepts have a very high communication volume, which leads to long transaction times and high bus loading. Thus, they are unsuitable for strictly real-time systems. A more effective way of realizing an atomic broadcast system is a so-called negative confirmation method. Here, in the fault-free case the bandwidth of the communication medium is not loaded by any acknowledgment traffic. Instead, in case of an error the faulty message is destroyed by the detecting node during the transmission. As a consequence, all nodes discard the faulty message and thus, data consistency is assured. This principle is used by the communication protocol CAN, which will be described later on.

If inconsistencies cannot be avoided it is necessary to detect them with a high probability (high error coverage), and with a minimal delay (low error latency) in order to be able to start fault tolerance measures. Component faults are to be tolerated through the use of redundancy to avoid a single point of failure leading to a breakdown of the entire system. For example, bus redundancy has to be provided for in order to maintain overall communication in spite of a faulty transmission channel. However, redundancy requires measures for error detection and localization, error notification, prevention of error propagation, redundancy switch-over, and the recovery of a consistent system state.

In the area of distributed control available bus systems do not supply these features sufficiently, in particular under strictly real-time constraints. The primary goal of fault-tolerant system design is that redundancy switch-over takes place without any loss, corruption, and duplication of messages. Therefore, error detection is one of the most important aspects. Concerning error detection, contemporary bus systems use so-called life-guarding methods. Life guarding is used to refer to the cyclical transmission of life messages by all of the operational nodes to a master node responsible for the error management. If a life message of a node does not occur, that indicates a component fault within that node. Depending on the cycle time of the life messages, an unacceptably long time may pass until the node failure is detected so that a loss of messages occurs leading to inconsistencies. In addition, time-driven protocols most often use time-out mechanisms to detect a node loss leading to unallowable fault latencies, too.

In addition, after the detection of errors all the other network nodes have to be informed of those inconsistencies within a minimum period of time. Otherwise, lost, corrupted, or duplicated messages cause further inconsistencies demanding comprehensive recovery operations. Thus, rapid error detection and notification mechanisms minimize the design effort for recovery measures and reduce fault tolerance latencies.

Fault tolerance comprises the separation of erroneous components in order to avoid the propagation of the fault to other system components. For example, a node which blockades the communication channel permanently by a short-circuit at a bus-side output has to be disconnected. Concerning distributed systems, the prevention of error propagation is often realized through the use of the fail-silent strategy (Kopetz and Grünsteindl, 1994): A fail-silent node is either operational as intended or do not produce any results at all.

Regarding event-triggered protocols, due to high error detection and notification latencies, the fail-silent strategy is not sufficient. Therefore, an active error mechanism has to be provided for detecting and indicating a component fault with a minimal delay. A realization of such a method, called fault-active mechanism, will be described later on.

3. CONCEPT OF A FAULT-TOLERANT DCCS

In the following, a distributed control system communication concept based on the above requirements and derived design rules is described.

3.1 System Architecture

As to be seen in Fig. 1, a DCCS includes several functional units such as process I/O, interfaces to other control levels, MMI (Man Machine Interface), process control, process monitoring and

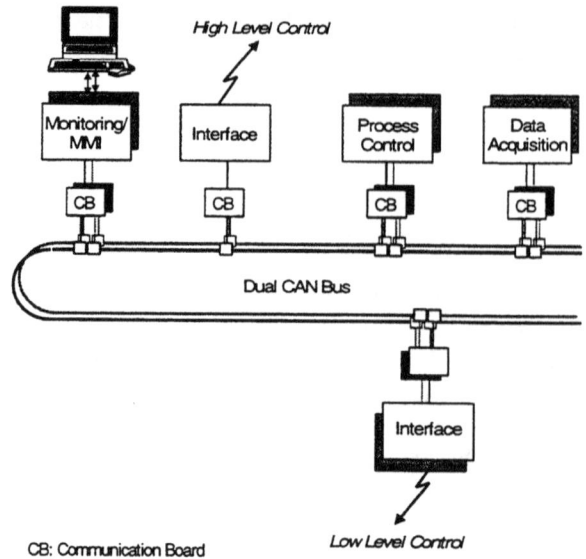

CB: Communication Board

Fig. 1. Sample DCCS Configuration

data acquisition. The configuration depends on the application domain as well as the target control level. According to the reliability requirements, a functional unit may be redundant. Each unit is attached to a dual CAN bus through the use of a communication board, also reffered to as node. A communication board provides two fully redundant bus links, each link comprising a microcontroller, a CAN communication controller, and a transceiver, as is illustrated in Fig. 2. The microcontroller initiates transmissions of data, selects received messages and passes them on for processing. In addition, the mechanisms for redundancy management and fault control are executed by the micro-controller. The CAN controller controls all of the mechanisms specified in the CAN protocol (*CAN Bus Specification 2.0*, 1994). Finally, the transceiver carries out the physical connection of the node to the transmission medium. Both electrical transmission media such as twisted pair or coaxial cables and optical waveguides can be used.

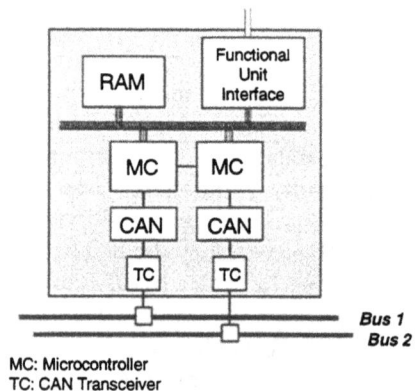

MC: Microcontroller
TC: CAN Transceiver

Fig. 2. Fault-Active Communication Board

42

3.2 *Communication Protocol*

The communication protocol provides effective measures for fault tolerance management, comprising error detection and notification as well as mechanisms for redundancy switch-over. For this purpose, the built-in features of the CAN protocol are enhanced.

3.2.1. *Basics of the CAN Protocol*

CAN, which was originally developed for use as a sensor/actuator bus in a motor vehicle, is suitable for wide areas of automation technology. Above all, the realization of a highly reliable and effective atomic multicast transmission principle and the mechanisms for fault detection make CAN a basis for highly reliable real-time systems.

CAN networks are multi-master systems, i.e. bus access for every node is allowed if the bus is idle, also referred to as CSMA/CR or Carrier Sense, Multiple Access, with Collision Resolution. A collision situation arises when two or more nodes try to transmit simultaneously. In this case, the message of highest priority prevails and may continue transmission. Therefore, a priority is assigned to every message. Thus, worst case access times for all messages can be calculated during system design (Tindell and Burns, 1994). The maximum data rate for a CAN system is specified at 1 Mbit/s.

The transfer of data in a CAN network takes place as follows. A node broadcasts a message to all other nodes. If any node discovers a transmission fault, it destroys the message while the transmission is still taking place by overwriting the bus level with an error frame. As a result, all of the nodes discard the erroneous message and the transmitter starts a new transmission. This corresponds to the principle of atomic multicast. A negative confirmation mechanism is realized, i.e. in a fault-free case, a transaction requires the transmission of only one message. Due to this as well as a very high protocol efficiency, CAN provides a very effective realization of an atomic multicast method.

CAN has built-in mechanisms for error detection and the localization of error sources. Defective nodes are switched off the bus, implementing a fail-safe behavior. For this purpose, each node maintains an internal error counter, which is incremented after a transmission fault is detected and decremented after each fault-free transmission. If an error counter reaches the value 127, the node automatically goes from the error active into the error passive state, i.e. it may continue to transmit and receive messages but cannot destroy any faulty message by transmitting an error frame.

A CAN controller having an error counter which reaches the value 256 switches into the bus off state, i.e. it no longer participates in the bus traffic in any way. That means, a faulty node disrupts the bus traffic by transmitting error frames until it goes into the passive state.

3.2.2. *Fault tolerance enhancements*

Since CAN is designed for use in single bus systems it has a number of disadvantages with respect to the requirements explained relating to fault tolerance. The limitations of CAN are:

- The management of redundancy is not provided for in the CAN protocol.
- Concerning node component faults, high error latencies may occur.

The redundant realization of system components, in particular of the bus line, requires additional mechanisms for the consistent switching over to redundant components in the case of a fault without exceeding real-time constraints. In addition, the negative confirmation mechanism of CAN causes high error latencies: the transmitter of a message does not detect the failure of another network node but rather assumes that if error frames do not occur all of the receivers have received its message without faults. Thus, node losses cannot be detected, especially in a sufficient period of time. The proposed protocol enhancements eliminate this through the use of a fault-active mechanism comprising system monitoring and error notification methods. The fault-active mechanism allows each communication link to be monitored by the other link of its node, as is illustrated in Fig. 2. For this purpose each micro-controller serves as a watchdog processor for the other link. In addition, in case of a component fault a node is able to become active autonomously, i.e. it informs all network nodes about the failure by transmitting an error notification message through the operational link and communication channel. Thus, the second bus system fulfills the function of a watchdog bus. During normal operation the entire process data traffic is executed through one communication channel. Fig. 3 shows a sample fault reaction process started after the occurrence of a failure of a CAN controller, which belongs to the bus system transmitting the process data:

(1) CAN controller a1 fails
(2) CAN controller a1 disrupts all network traffic on bus 1 until its error counter reaches 128
(3) reaching the error counter value 96 CAN controller a1 transmits an error interrupt to micro-controller a1
(4) micro-controller a1 informs micro-controller a2 of the loss of the CAN controller

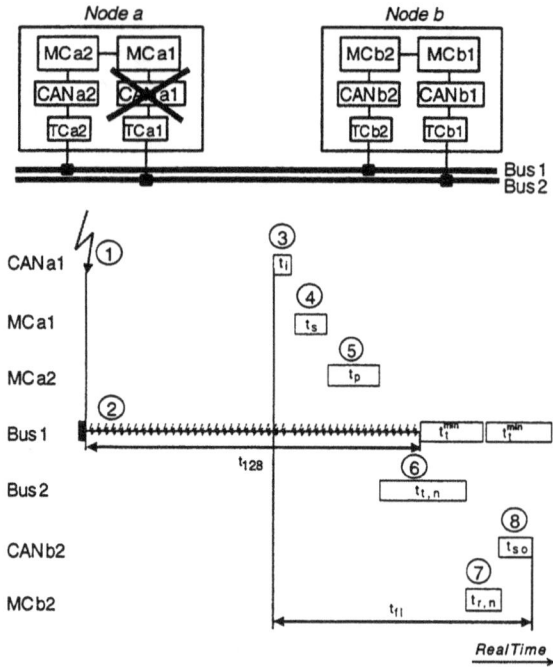

Fig. 3. Timing Diagram of a Fault Reaction Process

(5) micro-controller a2 starts the transmission of an error notification message through bus 2

(6) bus 2 transmits the error notification message

(7) all of the network nodes receive the error message through the links 2

(8) all nodes switch off bus 1 and continue the transmission of process data through bus 2

The advantage of this method lies in the fact that the fault tolerance process is executed while the faulty CAN controller is in the active error state (error counter ≤ 127). This controller therefore continuously destroys the message which is detected as being faulty up to the switch-over process. As a result, no message is lost and no faulty message is processed until the bus switch-over has finished. Concerning the recovery of a consistent system state after the occurrence of faults, an important aspect is the maximum number of messages, which are lost while the fault tolerance measures take place. Message losses may occur because the faulty node is not able to receive any message until it is replaced by its redundant component. Lost messages have to be retransmitted through the use of time-consuming recovery processes. Regarding the proposed concept, the worst case number of lost messages during a bus switch-over can be calculated as:

$$n \leq \left\lfloor \frac{t_{fl} - t_{128} - t_t^{min}}{t_t^{min}} \right\rfloor \quad (2)$$

$$= \left\lfloor \frac{t_i + t_s + t_p + t_{t,n} + t_{r,n} + t_{so} - t_{128} - t_t^{min}}{t_t^{min}} \right\rfloor$$

where:

n : is the maximum number of lost messages
t_{fl} : is the fault latency
t_{128} : is the error passive delay
t_t^{min} : is the minimum transmission delay
t_i : is the interrupt latency
t_s : is the error signaling delay
t_p : is the transmission processing delay
$t_{t,n}$: is the notification transmission delay
$t_{r,n}$: is the notification reception delay
t_{so} : is the switch-over delay

In case of a CAN controller fault no message loss occurs. Regarding specific malfunctions of a micro-controller, it may not be possible to entirely exclude a message loss. This situation can occur, for example, if the fault latency is greater than the duration of the transmission of a message. Nevertheless, only a few messages are to be retransmitted causing a minimal recovery effort. Thus, the preconditions of maintaining the consistency of data in the case of a fault are fulfilled.

4. CONCLUSIONS

Modern distributed control systems have to ensure both real-time operation and high fault tolerance. Concerning real-time behavior, a communication protocol adapted to the characteristic of data occurrence has to be provided for. Using standardized protocols, additional measures have to be taken with regard to fault tolerance aspects such as error detection and redundancy management. The underlying idea is to maintain data consistency even in the case of component faults without exceeding the timing constraints. The proposed system concept is based on an active error strategy using the built-in fault tolerance features of the standardized communication protocol CAN enhanced by redundancy management mechanisms. As a result, the technical demands can be fulfilled with minimal design effort.

5. REFERENCES

CAN Bus Specification 2.0 (1994). (12nc 9398 706 64011). Philips Semiconductors. Hamburg.

Kopetz, H. (1995). A Communication Infrastructure for a Fault Tolerant Distributed Real-Time System. In: *Proc. of DCCS'95*. Academic Press. Toledo. pp. 19–24.

Kopetz, H. and G. Grünsteindl (1994). TTP — a Protocol for Fault–Tolerant Real–time Systems. *IEEE Computer* pp. 14–23.

Tindell, K. and A. Burns (1994). Guaranteed Message Latencies for Distributed Safety Critical Hard Real–Time Networks. Ycs 229. Dept. Computer Science. Univ. of York.

DERIVATION OF RELIABILITY BOUNDS FOR REAL-TIME CONTROL SYSTEMS USING THE PATH-SPACE APPROACH

Jaeyoung Yoon, Daehyun Lee, Hagbae Kim, and Kwang Bang Woo *

* Department of Electrical Engineering
Yonsei University, Seoul, KOREA

Abstract: This paper proposes a reliability modeling of a real-time system embedded with the *hard deadline*, which implicitly represents the controlled process's needs in a form of timing information. Specifically, two types of tasks, clocked-based and event-based, of a real-time system are considered. The former utilizes a simple discrete-time Markov chain model extended with extra states describing the hard deadline as an integer multiple of the sampling period, whereas the latter constructs a semi-Markov model due to complicated nature of the deadline and the reliability bounds are derived by using the *path-space approach* to be especially appropriate for highly reliable systems which must recover from controller failures quickly.

Keywords: Reliability, real-time system, deadline, path-space, semi-Markov

1. INTRODUCTION

A real-time system is characterized by the property that the performance and/or the reliability of the system depends on logical correctness as well as response time of its behaviors. A common feature of those real-time control systems and embedded computers is that the controller-computers synergically interplay with the environments (including the plants).

However, most conventional reliability models/tools have concentrated on the separate behaviors of either controller computers or the plants (Butler and White, 1990; Somani and Sarnaik, 1990). Some assumed that the controller-computers must always be failure-free safely enough to control the plants, and others focused only on the states of the fault-tolerant computers considering temporary controller-failures as system failures. It is true in the real world that a controller failure may occur its rareness due to adopted fault-tolerant schemes and the system can survive temporarily in the presence of those controller failures. All these aspects should be captured in evaluating system reliability, especially in safety-critical systems such as aircraft or nuclear reactors. Since the time gap between controller failures and system failure can be described by the hard deadline, which is the implicit relation between the controller computer and the plant and thus formalizes the controlled process's needs in a form of timing information, a relibility modeling of a real-time control system should contain information about the hard deadline in a certain manner.

As previous work including the hard deadline into reliability modeling, the authors of (Shin and Krishna, 1986) used a Markov model not only to describe the component-failure behaviors but also to incorporate the deadline violation as a simple transition, capturing system failure caused by missing the deadline as well as exhausting spares. However, for the former, only the probability of missing a deadline was included into a simple transition. Other approach considered non failure-critical cases where some system-down time could be tolerated but the system-down state should be recovered within a certain deadline, where derived the mean value of the system lifetime and the cumulative operational time for the case of bounded repair time (Goyal et al., 1987). However, it is not trivial (computationally intractable) to derive the distribution from the Laplace-Stieltjes transform of the system lifetime.

In contrast to the work cited above, our model takes a global view of the hard deadline, and extend traditional reliability models, which generally consist of a normal (system-up) state and a system-down state governed by two basic events, fault occurrence and fault-/error- handling, by including the deadline information. We consider two types of behaviors, clocked-based and event-based, of the controllers in real-time systems, for both of which reliability will be evaluated with the hard deadlines. In discrete and periodic mode, the hard deadlines are simply derived as an integer multiple of the sampling time, and system reliability can be obtained by simple discrete-time Markov chain model with extra states accounting for the deadline. In continuous or aperiodic mode, system reliability can be modeled by using a semi-Markov model due to complicated nature of the deadlines, where the computation complication of integral equations (for the semi-Markov model) can be solved by introducing a new method, path-space approach, to yield the upper and the lower bounds of system reliability. Some numerical examples are presented to validate the effectiveness of our results.

2. RELIABILITY MODELING FOR CLOCK-BASED TASKS

In real-time control systems, the controllers are implemented with digital computers in the feedback loop, and A/D and D/A converters due to fast, accurate, and consistent performance as well as the capability and reliability of digital computers (Shin and Kim, 1992). The control input is computed by the controller computer at regular time intervals (e.g., the sampling period, T_s) without allowing job pipelining. This is classified as the clock-based tasks in a real-time system. The control inputs may not be correctly provided during consecutive sampling periods mainly due to the computer-controller failure(s)[1] When this period of malfunctions exceeds the hard deadline, the entire system will be led to catastrophe.

Since we distinguish the controller failure and the whole system failure, we need at least three states in a semi-Markov model to describe the behaviors of a dynamically-changing system, as depicted in Fig. 1-(a). (Note that this differs from traditional

---

[1] EMI is likely to induce transient computer failures (Kim and Shin, 1994).

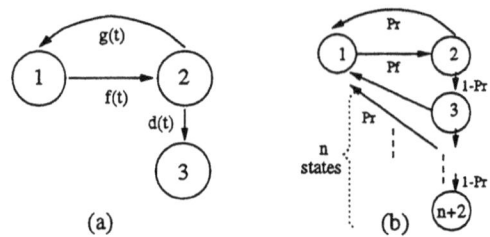

Fig. 1. An Markov-chain reliability model with the hard deadline for clock-based tasks

reliability models, basically composed of two different types of events; fault occurrence and fault-/error- handling.) Suppose that a clock-based task has a hard deadline D, equal to nT_s, which is also assumed to be deterministic, implying that no system failure occurs if the faults inducing computer failures disappear (or are recovered by a fault-tolerance mechanism) within nT_s. Then, the reliability model is built by changing the semi-Markov model into a discrete-time Markov chain model with $(n+2)$ states, as shown in Fig. 1-(b). In the modified model, we argue that (i) the additional n states account for the system resilience in the state of failed controllers, i.e., system failure results only if there are n consecutive incorrect (missing the update of) outputs of the controller computer, and (ii) the transition probabilities are readily calculated by using both the probability density functions of failure behaviors, $f(t)$ and $g(t)$, and the sampling period as follows:

$$p_f = \int_0^{T_s} f(t)dt \quad \text{and} \quad p_r = \int_0^{T_s} g(t)dt. \quad (1)$$

Then, the probability of system failure is easily computed using the transition probabilities and the probability of an initial state by means of simple matrix manipulation.

3. RELIABILITY MODELING FOR EVENT-BASED TASKS

There are a number of systems where actions have to be operated not at particular times or time intervals but in response to some events. These systems/tasks are classified as an event-based ones extensively indicating alarm conditions and triggering alarm actions in real-time systems. This kind of systems should respond within an allowable maximum time to a particular event. Events such that computer failures occur at non-deterministic intervals are referred as aperiodic ones having the hard deadlines. The reliability of the event-based system, where temporary controller-failures and entire system failure are distinguished as well, can be modeled by using a semi-Markov model like Fig. 1-(a).

We first discuss the assumptions required for a semi-Markov formulation of the hard deadline problem in an event-based system, and present a semi-Markov model and its Chapman-Kolmogorov equations to solve the state propobabilities. We then describe the path space approach to derive the upper and lower bounds for the probability of system failure due to a lengthy controller-down state. The final subsection contains numerical examples to validate the analytic results.

3.1 Semi-Markov formulation

We begin with a semi-Markov model using the assumptions:

- recovery is perfect, i.e., *as-good-as-new*, if done in time
- recovery distribution depends only on elapsed time since malfunction occurrence (independence on the global time)
- deadline is some fixed time (deterministic)
- all the processes (malfunction, recovery, and deadlines) are independent

Let $f(t)$, $g(t)$, and $d(t)$ be the density function for the arrival of the malfunction, the density for recovery, and the density for the deadline, respectively. With those four assumptions a semi-Markov model with three states, as given in Fig. 1-(a), is adopted from now on.

The first assumption that recovery is as-good-as-new is appropriate if the entire unit is replaced or reconfigured, and it is a good approximation for the repair of high quality equipment, such as the electronic components of a controller, during its useful life. The second assumption says that the repair procedure begins when a breakdown occurs, and that the repair time distribution remains the same throughout the lifetime of the system. (It is also models the steady state of a human repair crew that becomes neither much more efficient nor highly fatigued.) The assumption that the deadline is a fixed time is made for initial convenience. Once the formulas have been derived, a small modification extends the results to random hard deadlines in the future study. The fourth assumption reflects the fact that the deadline comes from external requirements, not the breakdown rate of the system or the efficiency of the recovery procedure.

To compute the probability of being in state 3 by time T given the system starts in state 1 at time 0. Chapman-Kolmogorov equations are:

$$P_{13}(t) = \int_0^t f(x) P_{23}(t-x) dx$$

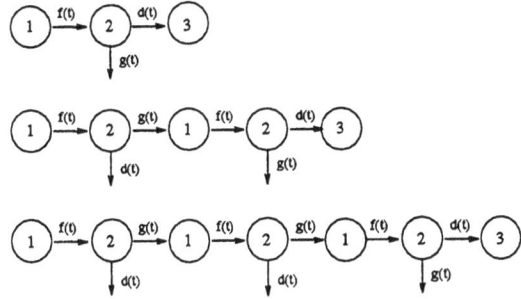

Fig. 2. The first three paths for the hard deadline model

$$P_{23}(t) = \int_0^t g(x)\bar{D}(x)P_{13}(t-x)dx + \int_0^t d(x)\bar{G}(x)dx \quad (2)$$

where $P_{ij}(t)$ is the probability of being in state j by time t given the system is in state i at time 0, and a bar above a distribution function indicates its complement. These recursive integral equations are analytically intractable except for a few simple cases. Furthermore, if the malfunction occurs at a slow rate while recovery and the deadline are fast, then the equations are numerically stiff.

3.2 Path-space approach

The path space approach considers all the ways of getting from the initial state to the absorbing state. If there are any loops in the model, then there are an infinite number of paths. The first three paths for the model in Section 2 appear in Fig. 2.

Each path is a disjoint event even though they share similar states. The probability of being in state 3 is the sum of the probabilities of traversing each of the paths. Hence, an upper bound for being in state 3 is the sum of the upper bounds for traversing each of the paths. (Similarly for the lower bound.) Initially the path space approach looks more complicated, but the regular and repetitive structure of the paths lets us derive useful formulas. Before deriving the bounds, note that since the density $d(t)$ is for a fixed time of length τ, we can define the probabilities of successful and unsuccessful recovery as:

$$Q = \int_0^\infty g(x)\bar{D}(x)dx = \int_0^\tau g(x)\bar{D}(x)dx$$

$$\bar{Q} = 1 - Q = \int_0^\infty d(x)\bar{G}(x)dx = \int_0^\tau d(x)\bar{G}(x)dx. \quad (3)$$

3.3 *Derivation of the bounds*

We now derive bounds for traversing the second path above. Derivation of the bounds for longer paths merely requires more bookkeeping. The probability of traversing the second path by time T is given by the convolution integral:

$$\int_0^T f(t_1) \int_0^{T-t_1} g(t_2)\bar{D}(t_2) \int_0^{T-t_1-t_2} f(t_3) \int_0^{T-t_1-t_2-t_3} d(t_4)\bar{G}(t_4)$$

$$dt_4\,dt_3\,dt_2\,dt_1 = \int_0^T f(t_1) \int_0^{T-t_1} f(t_2) \int_0^{T-t_1-t_2} g(t_3)\bar{D}(t_3)$$

$$\int_0^{T-t_1-t_2-t_3} d(t_4)\bar{G}(t_4)\,dt_4\,dt_3\,dt_2\,dt_1.$$

Adjusting the limits of integration gives an upper bound of:

$$\int_0^T f(t_1) \int_0^{T-t_1} f(t_2) \int_0^{\infty} g(t_3)\bar{D}(t_3) \int_0^{\infty} d(t_4)\bar{G}(t_4)$$

$$dt_4\,dt_3\,dt_2\,dt_1 = \int_0^T f(t_1) \int_0^{T-t_1} f(t_2)Q\bar{Q}\,dt_4\,dt_3\,dt_2\,dt_1,$$

and a lower bound of:

$$\int_0^{T-2\tau} f(t_1) \int_0^{T-2\tau-t_1} f(t_2) \int_0^{\tau} g(t_3)\bar{D}(t_3) \int_0^{\tau} d(t_4)\bar{G}(t_4)$$

$$dt_4\,dt_3\,dt_2\,dt_1 = \int_0^{T-2\tau} f(t_1) \int_0^{T-2\tau-t_1} f(t_2)Q\bar{Q}\,dt_4\,dt_3\,dt_2\,dt_1.$$

The two dimensional convolution integral that appears in both bounds is the probability that two or more events occur by time T. For the general case let $P[n \geq k|T]$ be the probability that k or more events have occurred by time T. An Upper Bound (UB) for being in state 3 of Fig. 1-(a) by time T is:

$$UB = \sum_{k=1}^{\infty} P[n \geq k|T]Q^{k-1}\bar{Q}, \qquad (4)$$

and a Lower Bound (LB) is:

$$LB = \sum_{k=1}^{\lfloor T/\tau \rfloor} P[n \geq k|T-k\tau]Q^{k-1}\bar{Q}. \qquad (5)$$

The upper limit of summation for the lower bound is the largest integer less than or equal to the quotient.

3.4 *Numerical examples*

This subsection considers a variety of examples: an assembly line where the reliability is moderate, an aircraft where the operating time is short and the reliability is high, and a satellite where the operating period is long and the reliability is fairly high. The parameter values are chosen to yield stress cases for the formulas (and may not reflect the ultimate in realism). Each examples uses both an exponential density $f_1(t) = \lambda e^{-\lambda t}$ and a gamma density $f_2(t) = \alpha^2 t e^{-\alpha t}$ for the occurrence of malfunctions. For each example, the parameters are chosen so that the exponential and gamma distributions have the same mean, which means $\alpha = 2\lambda$. For the assembly line, the operating period is one day, the hard deadline is fifteen minutes, the probability of successful recovery is 0.95, and the expected time to malfunction is ten days. For the exponential distribution:

$$LB = 0.004931, \qquad UB = 0.004988$$

$$Rel\ Error = (UB - LB)/UB = 0.01134.$$

For the gamma distribution.

$$LB = 0.000879, \qquad UB = 0.000862$$

$$Rel\ Error = (UB - LB)/UB = 0.01954.$$

For the aircraft with a highly reliable redundant/reconfigurable controller, the operating time is one hour, the hard deadline is one second, the probability of recovery is 0.99, and the expected time to malfunction is 10^8 hours. For the exponential distribution:

$$LB = 5.00000 \times 10^{-10}, \qquad UB = 4.99974 \times 10^{-10},$$

$$Rel\ Error = (UB - LB)/UB = 5.5525 \times 10^{-5}.$$

For the gamma distribution,

$$LB = 4.99600 \times 10^{-17}, \qquad UB = 4.99600 \times 10^{-17}.$$

This example demonstrates that the path-space approach with its upper and lower bounds are suitable for extremely stiff problems.

For the satellite with along mission time and reliable controllers, let the operating time be 20,000 hours, the hard deadline be thirty minutes, the probability of successful recover be 0.97, and the expected time to malfunction be 10^4 hours. For the exponential distribution:

$$LB = 0.058236, \qquad UB = 0.058236,$$

$$Rel\ Error = (UB - LB)/UB = 7.5048 \times 10^{-7}.$$

For the gamma distribution,

$$LB = 0.051445, \qquad UB = 0.051442,$$

$$Rel\ Error = (UB - LB)/UB = 6.8839 \times 10^{-5}.$$

4. SPECIAL CASES IN THE PATH-SPACE APPROACH

This section includes (i) an analytic upper bound for the exponential distribution as a special case of the previous section, (ii) a rough-and ready upper bound based on the expected number of interruptions, and (iii) a state aggregation example.

4.1 *Analytic upper bound for the exponential distribution*

For an exponential distribution with parameter λ the probability of exactly k events in time T is $P_k = (\lambda T)^k e^{-\lambda T}/k!$. Hence, the upper bound:

$$\sum_{k=1}^{\infty} P[n \geq k|T]Q^{k-1}\bar{Q}$$

can be displayed as the sum of $\bar{Q}$ times the rows.

$$
\begin{array}{lllll}
1 & -P_0 & & & \\
Q & -QP_0 & -QP_1 & & \\
Q^2 & -Q^2P_0 & -Q^2P_1 & -Q^2P_2 & \\
Q^3 & -Q^3P_0 & -Q^3P_1 & -Q^3P_2 & -Q^3P_3 \\
\vdots & \vdots & \vdots & \vdots & \vdots
\end{array}
$$

Temporarily ignore the first column. The top diagonal is:

$$\sum_{i=0}^{\infty} Q^i P_i = \sum_{i=0}^{\infty} \frac{Q^i (\lambda T)^i e^{-\lambda T}}{i!} = -e^{-\lambda T} e^{Q\lambda T}.$$

Each diagonal is Q times the diagonal above. Hence the sum of all the diagonals is:

$$-\sum_{j=0}^{\infty} Q^j [e^{-\lambda T} e^{Q\lambda T}] = -\frac{e^{-\lambda T} e^{Q\lambda T}}{1-Q}.$$

Adding the sum of the first column gives:

$$\frac{1}{1-Q} - \frac{e^{-\lambda T} e^{Q\lambda T}}{1-Q}.$$

Multiplying by $\bar{Q} = 1 - Q$ gives an upper bound of $1 - e^{-\lambda T\bar{Q}}$.

Summing the terms in a different order gives the correct answer since the series converges absolutely

4.2 *Rough and ready upper bound*

When $\lambda T\bar{Q}$ is small, an approximation to the analytic upper bound for the exponential is $1 - e^{-\lambda T\bar{Q}} \lambda T\bar{Q}$, which is $\bar{Q}$ times the expected number

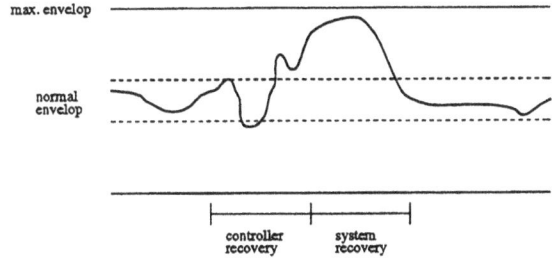

Fig. 3. Control scenario with two envelops

of events. In general, if Q is close to one, the upper bound

$$\sum_{k=1}^{\infty} P[n \geq k|T]Q^{k-1}\bar{Q} \approx \bar{Q} \sum_{k=1}^{\infty} P[n \geq k|T],$$

which, once again, is $\bar{Q}$ times the expected number of events.

4.3 *State aggregation example*

The three state model in Fig. 1-(a) is more general than it appears because of the technique of state aggregation in semi-Markov models. As an example considers the control scenario where there is a normal operating envelop inside a safe operating envelop. This is illustrated in Fig. 3.

In this scenario the system is vulnerable even after the controller has recovered because the system is close to the edge of its maximum safe envelop. A worst case analysis assumes that the system needs time for the controller to bring it within its normal operating envelop before it can survive another malfunction. A simple approach considers two fixed time intervals — (i) the maximum time the controller has to recover before the system leaves the outer envelop given it begins within the inner envelop and (ii) the minimum time needed by the controller to bring the system within the inner envelop given the system is inside the outer envelop. Given these two time intervals, a semi-Markov model is displayed in Fig. 4, where densities $h(t)$ and $v(t)$ are fixed time jumps of length τ_1 and τ_2, respectively. The densities $f(t)$ and $u(t)$ describe malfunction occurrence and the recovery procedure. States 2 and 4 inside the dashed box will be combined into a single state. To begin this process let:

$$Q_1 = \int_0^{\infty} u(t)\bar{H}(t)dt = \int_0^{\tau_1} u(t)\bar{H}(t)dt$$

$$Q_2 = \int_0^{\infty} v(t)\bar{F}(t)dt = \int_0^{\tau_2} v(t)\bar{F}(t)dt,$$

and let

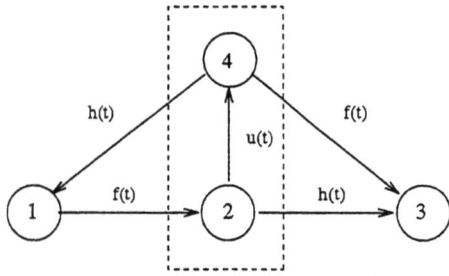

Fig. 4. Semi-Markov model for a control scenario

$$G(t) = \int_0^t u(t_1)\bar{H}(t_1) \int_0^{t-t_1} v(t_2)\bar{F}(t_2)dt_2 dt_1 = \int_0^t g(x)dx.$$

Then:

$$Q = G(\infty) = \int_0^\infty u(t_1)\bar{H}(t_1) \int_0^\infty v(t_2)\bar{F}(t_2)dt_2 dt_1$$

$$= \int_0^{\tau_1} u(t_1)\bar{H}(t_1) \int_0^{\tau_2} v(t_2)\bar{F}(t_2)dt_2 dt_1 = Q_1 Q_2$$

$$= \int_0^{\tau_1} u(t_1)\bar{H}(t_1) \int_0^{\tau_2} v(t_2)\bar{F}(t_2)dt_2 dt_1 + 0 + 0$$

$$= \int_0^{\tau_1} u(t_1)\bar{H}(t_1) \int_0^{\tau_2} v(t_2)\bar{F}(t_2)dt_2 dt_1$$

$$+ \int_{\tau_1^+}^{\tau_1+\tau_2} u(t_1)\bar{H}(t_1) \int_0^{\tau_1+\tau_2-t_1} v(t_2)\bar{F}(t_2)dt_2 dt_1$$

$$+ \int_0^{\tau_1} u(t_1)\bar{H}(t_1) \int_{\tau_2^+}^{\tau_1+\tau_2-t_1} v(t_2)\bar{F}(t_2)dt_2 dt_1$$

$$= \int_0^{\tau_1+\tau_2} u(t_1)\bar{H}(t_1) \int_0^{\tau_1+\tau_2-t_1} v(t_2)\bar{F}(t_2)dt_2 dt_1$$

$$= G(\tau_1 + \tau_2).$$

To continue the state aggregation process let

$$D(t) = \int_0^t h(t_1)\bar{U}(t_1)dt_1 + \int_0^t u(t_1)\bar{H}(t_1)$$

$$\int_0^{t-t_1} f(t_2)\bar{V}(t_2)dt_2 dt_1 = \int_0^t d(x)dx.$$

Then, as before,

$$D(\infty) = D(\tau_1 + \tau_2) = 1 - Q.$$

Hence, the four state model in Fig. 4 can be reduced to the three state model in Fig. 1 in a manner for which the derivation of the upper and lower bounds remains valid.

A general method for obtaining the probabilities and moments for state aggregation in semi-Markov models is developed in (White, 1992).

5. CONCLUSION

The hard deadline problem has not received much attention because it is more difficult to model and compute than the average throughput problem in reliability and performance analysis. This paper deals with two kinds of tasks, clock-based and event-based. A discrete-time Markov chain model with extra states accounting for the hard deadline is used to accurately derive system reliability for the former, whereas the latter adopts a semi-Markov model and takes a path space approach instead of the more familiar Chapman-Kolmogorov equations for the feasibility of computation. The upper and lower bounds that demand only simple parameters are derived for the probability of system failure. The path yields convenient formulas and straightforward computational techniques. Since deadline, complicated models can be reduced to a canonical simple model for analysis, which is shown in simple examples for the assumption of repair strategies — repair-as-good-as-new.

6. REFERENCES

Butler, R. W. and A. L. White (1990). SURE reliability analysis. *NASA Technical Paper*.

Goyal, A., V. Nocola, A. N. Tantawi and K. S. Trividi (1987). Reliability of systems with limited repairs. *IEEE Trans. on Reliability* **R-36**(2), 202–207.

Kim, H. and K. G. Shin (1994). Modeling externally-induced faults in controller computers. In: *Proc. 13rd IEEE/AIAA Digital Avionics Systems Conf.*. Phoenix, AZ. pp. 402–407.

Shin, K. G. and C.M. Krishna (1986). New performance measures for design and evaluation of real-time multiprocessors. *Int'l J. Computer Science & Engineering* **1**(4), 179–191.

Shin, K. G. and H. Kim (1992). Derivation and application of hard deadlines for real-time control systems. *IEEE Trans. on Systems, Man, and Cybernetics* **22**(6), 1403–1413.

Somani, A. K. and T. R. Sarnaik (1990). Reliability analysis techniques for complex multiple fault-tolerant computer architectures. *IEEE Trans. on Reliability* **39**(5), 547–556.

White, A. (1992). Simplifying semi-markov fault recovery models. In: *Proc. 11st IEEE/AIAA Digital Avionics Systems Conf.*

A CONTENTION-RESOLVING ALGORITHM FOR REAL-TIME COMMUNICATION BETWEEN DISTRIBUTED PROGRAMMABLE CONTROLLERS

Jung Woo Park, Hyeok Gi Park, and Wook Hyun Kwon

School of Electrical Engineering
Seoul National University, Seoul 151-742, KOREA
Tel: +82-2-873-2279, Fax : +82-2-878-8933
E-mail : jwpark@isltg.snu.ac.kr

Abstract: In this paper, a communication protocol is proposed for real-time distributed programmable controllers (PCs). Architecture for a real-time distributed system consisting of small PCs is also proposed. A basic execution cycle of the sequence control tasks executed by the PCs is defined, and a workload model which includes the execution period and time constraints is proposed. A contention-resolving algorithm is suggested in order to provide deterministic communication delays. Based on the workload model, the response time characteristic of the proposed communication protocol is investigated and compared with those of three well-known communication protocols. It is shown that the proposed real-time communication protocol provides better performance than do other protocols.

Keywords: Real-time communication; Contention-resolving algorithm; PC(Programmable Controller)

1. INTRODUCTION

Currently, programmable controllers (PCs) are often used in industrial control systems that require very elaborate controls (Roeder, 1984). As PCs take on more and more sophisticated tasks in a variety of fields, there is a strong need for real-time distributed architecture. Since the PC by its nature handles sequence control tasks in a real-time distributed PC system (RDPS), the overall sequence control task is broken down into smaller pieces which are then allocated to individual PCs. When the control tasks are allocated, data needs to be communicated among the PCs in the network (Schoeffler, 1984). The fundamental tasks of a PC consist of scanning input points, transferring the values of input points to other PCs, solving sequential and combinational logics, and updating output points (Warnock, 1988). A PC executes these four tasks repeatedly. Since the control outputs should be generated periodically, which is required in most control systems, the execution time of the above fundamental tasks must have a deterministic limit. But the communication delay which is required to transfer the values of input points to other PCs is not deterministic in most conventional PCs. The communication delay depends on the communication scheme used by PCs. There are three types of communication schemes for the sharing of a common communication medium (Bux, 1981). One is the token-passing technique, another is the random access technique, and the other is the master-slave technique.

Although contentions cannot occur when the token-passing technique is used, the communication delay is not deterministic. This is because the token arrival time is not periodic (Rubin, *et al.*, 1983). Thus, the token-passing technique is not appropriate to be used as a real-time communication scheme for the RDPS.

Contentions occur when the random access technique is used (Levi, *et al.*, 1990). Several contention-resolving algorithms have been studied (Capetanakis, 1979; Greene, *et al.*, 1981; Molle, *et al.*, 1985).As one of those algorithms, the tree conflict resolution protocol (Tree-CRP) utilizes statistical inference techniques to select the single message sender at a time slot (Capetanakis, 1979; Dechter, 1986).Although contention time is bounded, the tree conflict resolution algorithm can suffer deadlocks when the communication history information is unreliable (Molle, *et al.*, 1985). Other algorithms such as the virtual time CSMA/CD algorithm do not guarantee deterministic communication delay because they resolve contentions in a non-deterministic way(Molle, *et al.*, 1985).Thus, the random access technique cannot be used as a real-time communication scheme for the RDPS.

Contentions occur when the master-slave technique is used because the medium access time is determined independently by tasks. But the contentions can be resolved in a deterministic way if all the tasks are executed periodically. The contention-resolving algorithm suggested in this paper makes the master station detect the contention in advance by observing the execution periods and then resolves it by arranging the order of medium access. Thus, it guarantees deterministic communication delay even when contentions occur.

A detailed description of the contention-resolving algorithm is presented in this paper. The characteristics of the response time of the proposed real-time communication protocol with the contention-resolving algorithm are also analyzed.

This paper is organized as follows. Section 2 provides an overview of the distributed PC system. Section 3 defines the basic execution cycle of PCs, each of which has a intelligent processor. The contention-resolving communication algorithm is also suggested in this section. Section 4 provides a performance evaluation of the contention-resolving communication algorithm suggested in Section 3. The communication delay of the suggested algorithm is compared with those of other well-known protocols. Section 5 concludes this paper with possible extensions of this study.

2. THE RDPS ENVIRONMENTS

The PCs in the RDPS are interconnected via the bus topology which is widely used in the local area networks. Each PC has a local link memory. The purpose of the local link memory is to access remote information such as the status of input/output (I/O) points and the contents of registers of the other PCs in the network so that the PC can refer to this information to solve its own

sequential and combinational logics. The local link memory is divided into N blocks, where N is the number of nodes in the RDPS. One block of the local link memory, called a read/write (R/W) block in this paper, reflects the status of I/O points or the contents of registers in the corresponding PC.

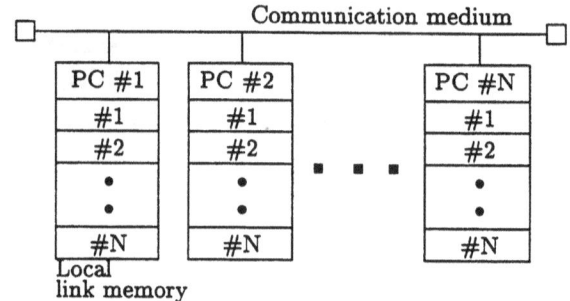

Figure 1: Distributed PC system

The other blocks of the local link memory, called read-only (R/O) blocks in this paper, contain replicated data of the other PCs in the RDPS. For instance, the block numbered one, #1 in Figure 1, contains local data of PC#1, and the blocks numbered two to N replicate those of the PC#2 to the PC#N respectively. That is to say, the R/W block of the PC#1 is the block #1, and the R/O blocks are the blocks #2 to #N in Figure 1.

The local link memory blocks that have identical block numbers should always contain identical data. Since the contents of a R/W block can be changed after scanning input points or solving sequential and combinational logics, the updates of the R/W block have to be transmitted to other PCs through the communication medium. In the real-time sequence control applications in the RDPS, contentions occur when two or more PCs begin to transmit the updates at the same time. The contentions can possibly defer time-constrained control operations if they are not resolved in a deterministic way. Thus, a contention-resolving algorithm is suggested to resolve the contentions among PCs in a deterministic way.

In order to justify the contention-resolving algorithm, the architecture of PCs has to be considered. Essentially, PCs have three functional units: the I/O unit (IOU), the logic-solving unit (LSU), and the network service unit (NSU). The IOU scans input points and updates output points; the LSU solves sequential and combinational logics; and the NSU updates the link memory.

The models in Table 1 are classified by the architecture of PCs in the RDPS. There are five combinations to implement a PC with one to three intelligent processors. Model 1 entrusts a main pro-

Table 1: Architectures of PCs

Model	Processor 1	Processor 2	Processor 3
1	LSU, IOU, NSU	X	X
2	LSU, NSU	IOU	X
3	IOU, NSU	LSU	X
4	LSU, IOU	NSU	X
5	LSU	IOU	NSU

cessor to handle the three functional units. Most small PCs might be consistent with this model. Models 2, 3 and 4 have two processors which divide the three functional units into two parts. One unit is dedicated to one processor and the other two are dedicated to the other processor. These models are often implemented as medium PCs. In large PCs, the three functional units are operated in parallel by three processors respectively.

The NSU functions independently in Models 4 and 5, while at least two functional units, including the NSU, are implemented in one intelligent processor in Models 1, 2 and 3 in Table 1. In the latter case, timely execution of the LSU and/or IOU may be affected by interruptions from other PCs in the network. Consequently, the control outputs cannot always be generated periodically. Thus, a communication protocol is needed in order not to interfere with the execution of the LSU and IOU. Since Model 1 represents the most constrained architecture, a communication protocol is studied for the architecture of this model. It is then easy to utilize the communication protocol for the other models.

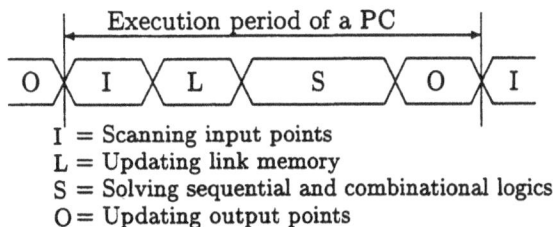

I = Scanning input points
L = Updating link memory
S = Solving sequential and combinational logics
O = Updating output points

Figure 2: Basic execution cycle

As shown in Figure 2, PCs execute four fundamental tasks in a repeated sequence. In the interval marked 'I' in Figure 2, a PC scans its own input points. In the interval marked 'L', the PC informs other PCs of its latest contents of the R/W block in the local link memory that holds status of I/O points or intermediate results derived from the previous execution period. In the interval marked 'S', the PC solves sequential and combinational logics with the status of its own input points and the contents of the local link memory. Then, in the interval marked 'O', the PC updates the output points.

In the case that the processor is in the interval

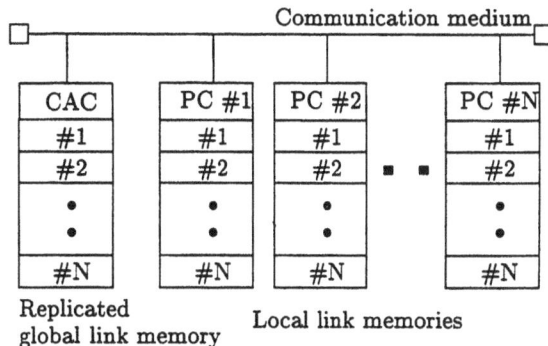

Figure 3: Suggested architecture of RDPS

marked 'S', interruptions by communication requests from other PCs may defer the solving of sequential and combinational logics. Then the PC cannot generate the control outputs periodically. Thus, a master node which is called the *Central Access Controller* (CAC) is introduced, as shown in Figure 3. The CAC node has the replicated global link memory and controls access to the shared communication medium. The replicated global link memory is dedicated to collect the latest updates of the R/W blocks in a PC and distribute the updates to other PCs.

3. THE CONTENTION-RESOLVING ALGORITHM

Since communication with other PCs occurs once in an execution period of a PC, as shown in Figure 2, it is possible to put those fundamental tasks into a workload. The workload J_k of the PC#k is defined by a four-tuple:

$$J_k = (P_k, \ EBT_k, \ LTT_k, \ ST_k).$$

A set of workloads is defined as:

$$\mathbf{J} = \{J_1, J_2, \cdots, J_k, \cdots, J_N\},$$

where N is the total number of PCs in RDPS. The execution period P_k of J_k is the time interval in which the four fundamental tasks have to be completed, as shown in Figure 2. The EBT_k is the time that the PC#k completes scanning the input points and begins updating the local link memory (where EBT is the *Earliest Begin Time*). The LTT_k of each task in any period of J_k is the time constraint before which a PC has to end its access to the communication medium (where LTT is the *Latest Terminating Time*). The time constraint LTT_k has to be met to keep the execution period P_k unchanged, and then to generate the control outputs periodically. The amount of time required to scan input points and update output points is constant because the number of I/O points in a PC is not variable. The logic-solving time in the execution period would vary according to application tasks, but it is limited by

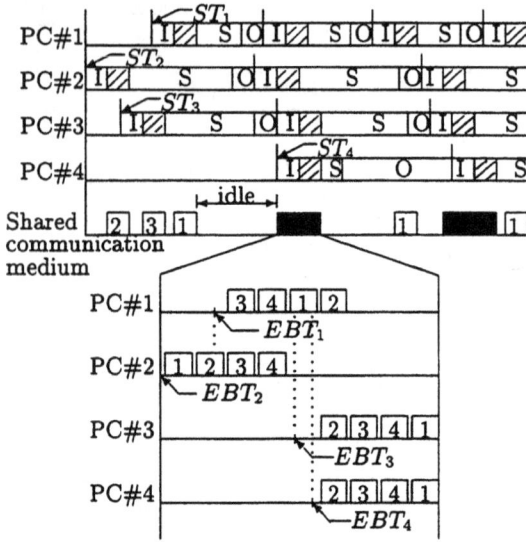

Figure 4: An example of medium access control

the worst-case logic-solving time. The ST_k represents the initial begin time of J_k (where ST is the *Start Time*).

Figure 4 shows an example in which a set of workloads, **J**, consists of four workloads. That is, there are four PCs in the example:

$$\mathbf{J} = \{J_1, J_2, J_3, J_4\},$$

where the workloads are defined as:

$$J_1 = (250, 53, 108, 133)$$
$$J_2 = (380, 40, 95, 0)$$
$$J_3 = (350, 46, 101, 64)$$
$$J_4 = (400, 43, 98, 423).$$

Allocation of the shared communication medium is illustrated in the centre of Figure 4. The number inside the small white rectangle on the line titled *'shared communication medium'* indicates the PC that has exclusive access rights to the shared communication medium at that time. Thus, the numbers correspond to the slashed rectangle respectively as shown in Figure 4.

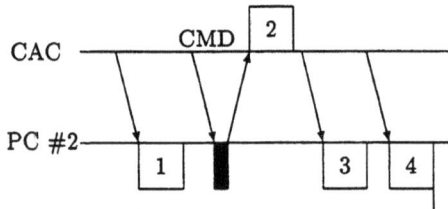

Figure 5: Updating single local link memory

Figure 5 illustrates the situation when only one PC (PC#2 in this example) begins to update its local link memory. The CMD on the line titled 'CAC' means the CAC node transmits a command packet rather than a data packet which contains the contents of a link memory block. The shaded rectangle beneath the line titled 'PC#2'

denotes that the PC#2 receives the command packet and prepares to transmit the contents of the R/W block. This is the sequence to update a single local link memory, which should be completed by the end of the given time constraint, LTT_k. And this time constraint should also be met when contentions occur as shown in the shaded interval in Figure 4.

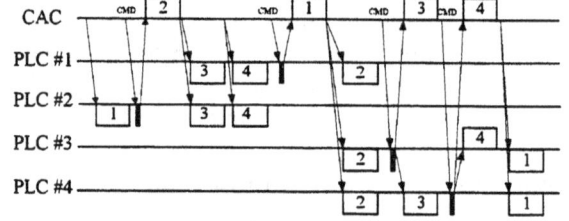

Figure 6: Updating multiple local link memories

Figure 6 illustrates the sequence where the PCs involved in the contentions update the local link memory by transmitting their R/W blocks and receiving their R/O blocks according to the contention-resolving algorithm proposed. The time interval in which the PC#2 updates its local link memory in Figure 6 has exactly same length as the example in Figure 5. Thus, from the viewpoint of individual PCs, it is clear that the sequence in which a PC updates its local link memory consists of receiving three R/O blocks and one command and transmitting one R/W block whether contentions occur or not.

The contention-resolving algorithm is described in Figure 7. The algorithm is executed by the CAC node. The contention-resolving algorithm maintains a vector *Earliest Access Time* EAT to indicate the earliest access times of the shared communication medium:

$$EAT = (EAT_1, EAT_2, ..., EAT_N),$$

where EAT_i is the earliest time when the PC#i tries to update its local link memory by accessing the shared communication medium.

Initial values of EAT_i for all i will be:

$$EAT_i \leftarrow ST_i + EBT_i,$$

where ST_i is the start time specified in the workload definition. The value of EAT_i is updated when the PC#i completes updating its local link memory. The *next* EAT_i will be:

$$EAT_i \leftarrow EAT_i + P_i.$$

The set S_{ja} contains the number of PCs that have arrived before transmission of local link memory blocks begins. The set S_{ja} is initially an empty set. The variable M_i counts the number of updated local link memory blocks of the PC#i to indicate whether the PC#i has updated all blocks of its local link memory. The variable B contains the link memory block number which is currently transmitted or received. The function C(t) is introduced to get current time.

Procedure UpdateLinkMemoryBlocks;
BEGIN
  $\forall i : EAT_i \leftarrow ST_i + EBT_i$;
  $S_{ja} \leftarrow \emptyset$;
Idle:
 repeat
  $\forall i : M_i \leftarrow 0$;
  $EAT_{min} = \text{MIN}(EAT_1, EAT_2, ..., EAT_N)$;
  $\forall i :$ **if** $(EAT_{min} = EAT_i)$ **then**
      $S_{ja} \leftarrow S_{ja} \cup \{J_i\}$;
      $M_i \leftarrow 1$;
    endif
  $B \leftarrow 1$;
 do
  nothing;
 while $(C(t) < EAT_{min})$;
 repeat
  if $(J_B \in S_{ja})$ **then**
   Let the PC#B broadcast the contents of the block #B;
   Receive and Store the contents of the block #B;
  else
   Broadcast the contents of the block #B;
  endif
  $t_1 \leftarrow C(t)$
  $\forall i :$ **if** $(EAT_i \leq t_i)$ *and* $(M_i = 0)$ **then**
      $S_{ja} \leftarrow S_{ja} \cup \{J_i\}$;
    endif
  $B \leftarrow B + 1$;
  if $(B > N)$ **then**
   $B \leftarrow 1$;
  endif
  $\forall J_i \in S_{ja} : M_i \leftarrow M_i + 1$;
  $\forall J_i \in S_m :$ **if** $(M_i > N)$ **then**
      $S_{ja} \leftarrow S_{ja} - \{J_i\}$
      $EAT_i \leftarrow EAT_i + P_i$;
      $M_i \leftarrow 0$;
     endif
  until $(S_{ja} \not\subset \emptyset)$
 until forever;
END;

Figure 7: The contention-resolving algorithm

With the contention-resolving algorithm, the execution periods of the PCs need not be prolonged even if all the PCs in the network require to access the shared communication medium at the same time. Furthermore, due to its simplicity, it is easy to implement the proposed contention-resolving algorithm with a microprocessor, and thus the PCs in the RDPS do not need an additional communication controller.

4. PERFORMANCE EVALUATION

There are two kinds of packets in the network, according to the algorithm proposed. One is the packet to transmit a command to each PC and the other is the packet to broadcast one block of the replicated global link memory in the CAC node or the local link memories in PCs. We take the length of command to be L_c bits while the header length is L_h bits, so that the communication delay is just $\mu_{cmd} = [L_h + L_c]/C$, where C is the cable transmission rate in bps.

Since each block of the local link memories and the replicated global link memory has equal length, the length of one block of link memory is simply L_M/N, where L_M is the total length of the local link memory in one PC, and N is the total number of PCs in the network. Thus, the communication delay to transmit one block of local link memory is defined as one block time: $\mu_{data} = [L_h + L_M/N]/C$.

It is shown in the previous section that the sequence in which a PC updates its local link memory consists of transmitting one R/W block and receiving $(N - 1)$ R/O blocks and one command message. Figure 4, however, shows that there exists an additional delay when the PCs involved in the contentions want to update their local link memories. As shown in Figure 4, PCs (except the PC#2 which terminates the idle period) cannot actually use the shared communication medium. But the additional delay is bounded to μ_{data} as shown in Figure 4. Hence, the worst-case time μ required to update the local link memory of a PC is:

$$\mu = N\mu_{cmd} + (N + 1)\mu_{data}.$$

The LTT_k of J_k defines the latest terminating time of the PC#k should terminate updating its local link memory. Thus, we can state that J_k is *feasible* if

$$LTT_k - EBT_k \geq \mu.$$

The workloads assigned to each PC are periodic. As shown in Figure 2, every execution period consists of four tasks. The response time of a PC can be defined as the execution time of those four tasks. The response time $\Re$ of J_k can be described as:

$$\Re = t_{input}[k] + t_{output}[k] + t_{solve}[k] + N\mu_{cmd} \\ + (N + 1)\mu_{data},$$

where $t_{input}[k]$ and $t_{output}[k]$ are input and output update times of the PC#k respectively.

The time spent to update input and output points is constant because the number of input and output points would not vary after the configuration of the PC was completed. The execution time to solve sequential and combinational logics can be altered by a conditional jump or program loop. But $t_{solve}[k]$ represents the worst-case execution time of the PC#k.

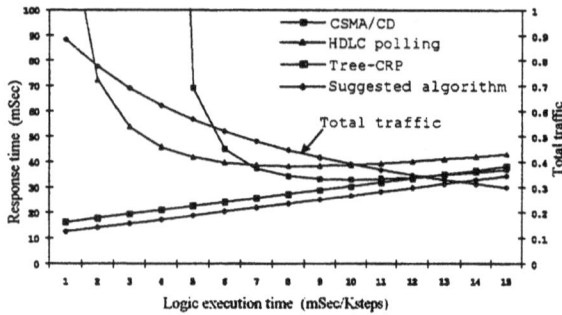

Figure 8: Comparison with other protocols

Figure 8 provides performance results for the case of four communication protocols operated at 1 Mbps. The number of PCs in the network has been chosen as $N = 16$. Figure 8 plots response time $\Re$ against the execution time of 1000 logic instructions. The curve (total traffic) is overlaid to show the comparative characteristic of the four communication protocols.

As the execution time of 1000 logic instructions increases, the period P_k of J_k is protracted. Thus, the total traffic of the network is decreasing. But the system response time $\Re$ is increased as the execution periods of individual PCs are prolonged. As it is known, the CSMA/CD scheme begins to fail when total traffic approaches 0.6. The walk time of the HDLC polling protocol is greatly increased at high traffic. The order of response time of the Tree-CRP strategy is actually same as that of the contention-resolving algorithm proposed in this paper, which is due to the fact that the Tree-CRP provides limited contention by a binary-search method. The difference in response time between two protocols is the time required to search one candidate to use the transmission medium exclusively. Although the Tree-CRP algorithm provides relatively good performance, it can not actually be used in the environment discussed in this paper. The programmable controllers discussed in this paper have only one intelligent processor so that they cannot co-operate the binary search algorithm in the Tree-CRP when they execute their programmed logic instructions. Thus, the suggested contention-resolving algorithm provides better performance than that of other protocols and satisfies the real-time constraints given to individual PCs.

5. CONCLUSIONS

In most control systems, the control outputs are required to be generated periodically. But in conventional distributed PCs it is very hard to generate the control outputs periodically because the communication delay is not deterministic. By using the proposed contention-resolving algorithm,

it becomes easy to generate the control outputs periodically. It is shown that the communication protocol with the contention-resolving algorithm provides better performance than do other protocols. Furthermore, due to simple architecture, it is easy to implement the proposed contention-resolving algorithm with a microprocessor, and thus the PCs in the RDPS do not need an additional communication controller.

There are special tasks such as interruptions and emergency tasks. It is very hard to meet the time constraints of these tasks, because they are not predictable. An algorithm to resolve the contentions invoked by those tasks will be studied in the future.

REFERENCES

Bux, W. (1981). "Local-Area Subnetworks: A Performance Comparison," *IEEE Tr. on Communications*, **Vol.COM-29**, pp.1465-1473.

Capetanakis J. I. (1979). "Generalized TDMA: The Multi-Accessing Tree Protocol," *IEEE Tr. on Communications*, **Vol. COM-27**, pp.1476-1484.

Dechter R. and Kleinrock L. (1986). "Broadcast Communications and Distributed Algorithms," *IEEE Tr. on Computers*, **Vol.C-35**, pp.210-218.

Greene E. P., and Ephremides A. (1981). "Distributed Reservation Control Protocols for Random Access Broadcasting Channels," *IEEE Tr. on Communications*, **Vol. COM-29**, pp.726-735.

Levi S. and Agrawala A. K. (1990). *Real-time System Design*, pp.238-253, McGraw-Hill Publishing Company.

Molle M. L. and Kleinrock L. (1985). "Virtual Time CSMA: Why Two Clocks are Better than One." *IEEE Tr. on Communications*, **Vol. COM-33(9)**, pp.919-933.

Roeder, W. H. (1984). "Communications Between Programmable Controllers in the Industrial Environment," *IEEE Tr. on Ind. Appl.*, **Vol.IA-20**, pp.504-509.

Rubin I. and De Moraes L. F. M. (1983). "Message Delay Analysis for Polling and Token Multiple-Access Schemes for Local Communication Networks," *IEEE J. Select. Areas Commun.*, **Vol. SAC-1**, No. 5, pp. 935-946.

Schoeffler, J. D. (1984). "Distributed Computer Systems for Industrial Process Control," *IEEE Computer*, pp.11-18.

Warnock Ian G. (1988). *Programmable Controllers, Operation and Application*, pp.60-80, Prentice Hall.

EASY IMPLEMENTATION OF DISTRIBUTED CONTROL SYSTEMS BY INTELLIGENT NODES AND AUTOMATIC CODE GENERATION

Jose A. Frutos-Redondo (*) and Jose M. Giron-Sierra (**)

*(*) Dept. Automatica, Escuela Universitaria Politecnica de Telecomunicaciones
Universidad de Alcala de Henares. 28871 Alcala de Henares, Madrid, Spain
e-mail: atfrutos@alcala.es Tf: (341) 885 48 32 ; Fax: (341) 885 48 04*

*(**) Dept. Informatica y Automatica, Fac. Fisicas
Universidad Complutense de Madrid. Ciudad Universitaria, 28040 Madrid, Spain
e-mail: gironsi@dia.ucm.es. Tf: (341) 394 43 87 ; Fax: (341) 394 46 87*

Abstract : The rapid prototyping and building of data acquisition and control systems is a need felt both by researchers and final users. The paper describes a modular system, based on intelligent nodes and the automatic, interactive, generation of code, that gives an easy solution for the creation of distributed systems.

Keywords : distributed control, programming environments, distributed artificial intelligence, computer interfaces.

1. INTRODUCTION

Influential researchers dealing with automatic control issues, point out the need of tools for the rapid prototyping of control systems (Arzen, 1993; Auslander, 1993). Such kind of tools will empower the study of different alternatives, and the quick implementation of a guaranteed solution. A valuable feature in this context, is the automatic generation of code (Marwedel and Goosens, 1995).

An important part of the current production, services, and research activities, is due to the contribution of small companies or groups, or even individuals. Many times, they do not dispose of technical knowledge about the control systems they need. Actually, we have observed this phenomenon, helping people from other Departments in our university, to solve control experimental problems. Having identified this fact,
we started a research project, to provide the means for the (small but important) activity protagonists to be able to easily implement its own distributed control systems.

From various information sources, including on-the-field interviews, we realized that the control function more requested and understood by industries, when adopting computer-based new technologies, is monitoring, supervision (perhaps embodying local tasks of regulation or automation). But this requires the computer interface to the real world, and it is well known that the handling of sensors and the adaptation to analog information transmission, can lead to difficulties and problems: in particular, when going apart from established standard measurements.

2. OBJECTIVES OF THE RESEARCH

In view of the mentioned needs and the obstacles found by a great variety of users, we decided to develop a way of solution based on intelligent nodes, capable of sensorial and actuation local tasks, and connected by digital transmission to a computer in a control site (Kohn et al., 1995)(figure 1). In harmony with this distributed architecture, we also investigate the implementation of software tools to quickly and easily create applications, through the automatic generation of code.

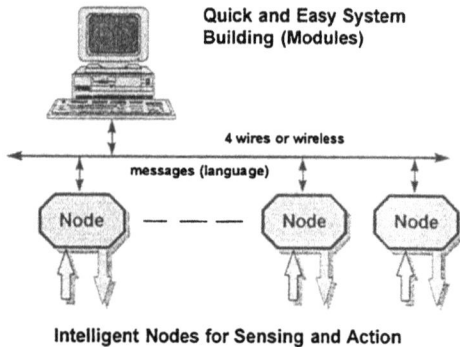

Quick and Easy System Building (Modules)

4 wires or wireless

messages (language)

Node — — — Node   Node

Intelligent Nodes for Sensing and Action

Figure 1 : architecture of the system

The sort of demands we are touching on, has been detected by different companies. So they try to cope with it, in a variety of ways. Many are intended to simplify the use of traditional hardware for data acquisition, such is the case of computer cards connected to sensors through analog signal conditioning electronics. In other direction, there is a lot of excitement concerning to fieldbuses, that may resort to the adaptation of methods and technology of conventional digital networks. Our idea is to create nodes with enough digital processing capability, so as to provide an infrastructure able to, for example, accomplish state and parameter estimation local functions ("software sensors").

The research line we are following, leads us to consider the distributed control issues from the perspective of cooperative agents (Maes, 1991; Jennings, et al., 1996), in a fashion similar to the multi-robotic systems designed to collaborate in a common task, by means of reactive behaviors and exchanges of messages.

In the following, we shall describe two coupled aspects of our system: the nodes, and the generation of code.

3. THE SYSTEM NODES

The core of each node is a Flashlite (JKmicrosystems) board. It is a small-size (10.7 x 9.2 cm.) PC for embedded applications. The processor is a NEC V25, 100% Intel compatible. The board includes 512 Kbytes of pseudo-static RAM, and 256 Kbytes of flash memory.

The software development is done on conventional MS-DOS computers. We employ Borland C++. When a program is ready for testing on the node, we compile it to EXE format, and download it to the Flashlite board, where it resides on the flash non-volatile memory (emulating a floppy disk). To start the execution we use the reset of the board.

To complete each node (figure 2), we add another board with analog and digital interface circuits, and digital data communication circuits. The analog interface is capable to be directly connected to a set of sensors (temperature, pH, pressure, etc.), to standard 4..20 mA instrumentation signals, and to 0..10 volts. signals. Four of the digital input channels are opto-isolated, and four of the digital output channels are connected to relays. The other digital channels use amplifiers.

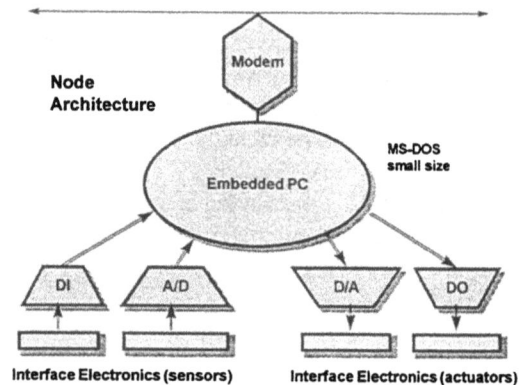

Node Architecture

Modem

Embedded PC

MS-DOS small size

DI   A/D   D/A   DO

Interface Electronics (sensors)   **Interface Electronics (actuators)**

Figure 2 : main blocks of the node

Following an object orientation, we define the class "node", which contains a set of processing and interface classes (figure 3). For instance, it disposes of the "controller" class, that may operate as a PID (or other alternatives, that are defined as discrete transfer functions). The "analog interface" class, that manages the signal conditioning and A/D conversion chips, use a set of objects: one for digital input/output, and for the analog signals as many as input and output channels. These classes act as intermediaries with regard to hardware, and present to the user a uniform functionality, independent of the technological solutions (the classes obey to a defined set of messages).

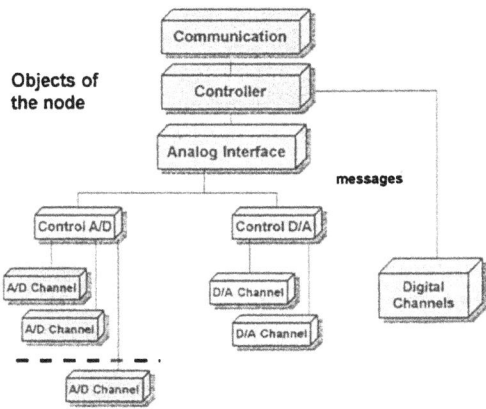

Figure 3 : classes and objects of the node

We have created a language for the exchange of orders and informations between nodes and the computer in the control site. Apart from the necessary orders for initialization and restarts, there are others for periodically testing the system, to see if the nodes work properly, and for alarms. The information exchange could be of a short or long nature: each node has buffers to maintain and analyze the trajectories of the input signals, and the control site could ask for a complete buffer contents. The "communication" class is in charge of the data transmission formats and protocols.

4. AUTOMATIC CODE GENERATION

We developed a software tool for the interactive building of applications on the computer. The main activity is the screen edition: the user selects windows from a set of options (figure 4). For instance, there are windows for the display of the trajectories of analog signals (each as measured by a selected node and channel). There are also windows to display the present status of analog (by numbers) or digital (by buttons) signals. With the mouse, the user sets up the position and size of these windows, mounting the application screens. Each window has a menu to send instructions to the related nodes (alarms settings, actuations, samplig periods, etc.), and to specify the functions of the window with respect to the nodes. When the screen edition is finished, the application can immediately begin to work: so it is easy to test the application as it is developed. When a satisfactory result is obtained, C++ code can be automatically generated. Using Borland C++, the code can be compiled, saved on a file, and directly executed.

The application so generated, talks to the nodes using the language we designed. In each node, the code of the class "node" is kept (non-volatile memory) to carry out the orders coming from the communication channel.

The result of the screen edition can be saved into intermediate code. This code may be used afterwards, to re-edit the screen and create a new application.

We employed the Quinn-Curtis libraries for the data visualization and specification dialog windows.

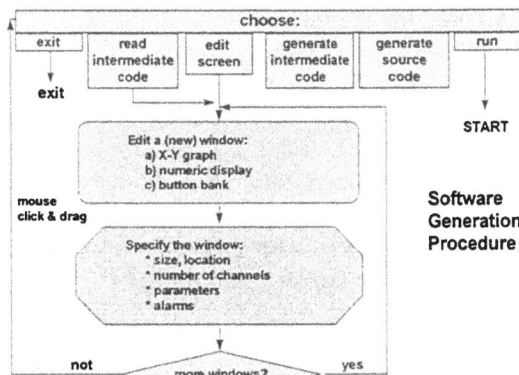

Figure 4 : automatic generation of code

5. SYSTEM CHARACTERISTICS

When the system is running, doing tasks of data acquisition and actuations, the software at the control site (as generated by the application generation tool) obeys to an object orientation similar to the nodes (figure 5), using a "communication" class, an "user interface" class that updates the screen and reads the keyboard, and a "control" class that manages the information flux from the nodes, and the orders (including user interventions).

Among the varied alternatives for digital intercommunication, we chose for the moment a solution via modem and cable. It is immediate to employ other supports (radio, optical fiber, etc.), for medium or long distances.

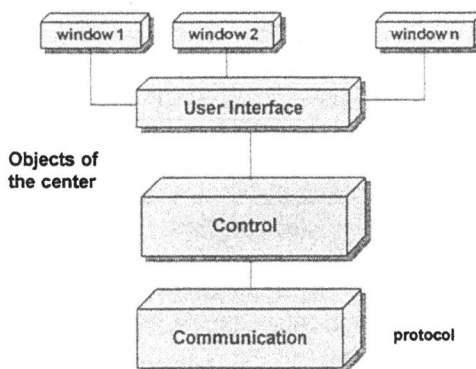

Figure 5 : classes and objects of the control site

The type of uses we are facing now (chemical processes, some manufacture examples), does not present hard difficulties of time response. Therefore, we devised a coordination of the system based on polling and an extra frequency for asynchronous high priority events.

As specified by the application, the nodes can detect when a signal breaks a limit (low or high): this means an alarm, and the control site will be informed.

One of the missions of the control site, is to detect malfunctions of the system, and to help to the user in the maintenance tasks.

The generated code is open: the user can edit and modify it to include special functions.

6. INTELLIGENT AGENTS

We consider each node as an intelligent agent. A lesson from expert systems history is that the most useful approach is to consider narrow knowledge subjects: that is, not many rules. This is favoured by intelligence distribution among specialists. In our case, an aspect forcing to confined expertise is that each node has a location in a plant, so it is connected to a concrete situation of expected signals characteristics.

During the application editing, the user can specify a set of rules for each node. The rules associate events (such are alarms) to messages to the control site, and actions on the plant. The node class includes a small rule-based decision class.

One of the facets we want to investigate, in the context of cooperative groups, is the autonomy degree of the nodes. Following the metaphor of a chief who instructs to new operators how to react to certain circumstances, several times, until the operators are trained and can be left to autonomous decisions; we let the nodes (the operators) take more and more autonomy. Each node is assigned an autonomy degree alpha (a number), by the control site. At the beginning, when a rule fires, the node sends a high priority message to the control site, and waits to the control site to confirm an action to take place. The node counts how many times a rule fires, and the action is confirmed. When the count surpass the number alpha, the node is allowed to apply the action without confirmation (but, still, it sends a low priority message to the control site). There are exceptions: decisions that always correspond to the control site.

In this way, the nodes come progressively to a reactive behavior, under user supervision.

The control site will listen to urgent messages, through the extra frequency. When there is one, the control site conducts a rapid polling, to establish a list of the alarms sent by the nodes. Having a general view of the problem detected, that can manifest several symptoms detected by some of the nodes, the control site will instruct to the nodes the correct reaction. If the problem means an important change of the plant behavior, the nodes are instructed not to be autonomous. Also, the problem could require the user intervention. All these functions are specified by the user, by means of a table of possible problems, with the symptoms, and the pertaining decisions.

During the normal polling, to acquire information from the nodes, the control site will be also informed of the decisions taken by the nodes.

7. RESULTS

The preliminary results we obtained with a first system, consisting of a PC and two nodes, are satisfactory indeed. We generate applications in few minutes. The training of users is fast. The nodes are compact and not expensive (under 400 $). We developed a set of software tools to help in the installation and maintenance tasks.

Regarding to the intelligent functions, we still need a lot of experimental work, with simulated and real plants, to test and refine our approach.

8. CONCLUSIONS

We have described a distributed system for data acquisition and control, with a modular architecture, based on intelligent nodes, and an automatic code generation tool. The main idea of the design, is to facilitate a quick and easy building of specific systems.

In the future we shall study the performances on different contexts: fermentations, gasoline synthesis, electromechanical systems, etc. With the acquired experience, we shall gradually enrich the intelligent functions of the nodes.

REFERENCES

Arzen, K.E. (1993). Using real time expert system for control system prototyping. *IEEE T. Sys. Man & Cyb.*, **v.2**, 25-30.

Auslander, D.M. (1993). Operator interface development for control software. *IEEE Control Sys. Mgz.,* August, 66-72.

Jennings, N.R. et al. (1996). Using ARCHON to develop real-world DAI applications, Part I. *IEEE Expert,* Dec., 64-70

Kohn, W. et al. (1995). A hybrid systems approach to computer-aided control engineering. *IEEE Control Sys. Mgz.,* April, 14-25.

Maes, P. (Ed.) (1991). *Designing autonomous agents.* MIT Press.

Marwedel, P. and G. Goosens (Eds.) (1995). *Code generation for embedded processors.* Kluwer Academic Publishers.

HOME AUTOMATION USING DCS TECHNOLOGY

James Or   Lin Qing   Dapeng Tien

Department of Electrical Engineering
Ngee Ann Polytechnic, Singapore
Email : olc2@np.ac.sg

Abstract: The paper presents a home automation model developed using DCS technology. The model has nodes that include smoke detector, door-contact sensor, door release, temperature sensor, air-cooler, motion detector, alarm panel, curtain control motor, brightness sensor, lighting control and hand-held IR controls. This paper also presents the software aspect which uses the NEURON C and how the nodes communicate with one another using network variable propagation. The network protocol used is also briefly discussed.

Keywords : Distributed Control System, Home Automation, LONWORKS, Neuron C, Neuron Chip, Local Operation Network, Predictive P-Persistent CSMA/CD Protocol.

1.   INTRODUCTION

In a centralised control system, all sensors and devices are connected individually to the micro-controller in the same way the computers are connected to the host computer. A typical block diagram of such system is as shown in figure 1.

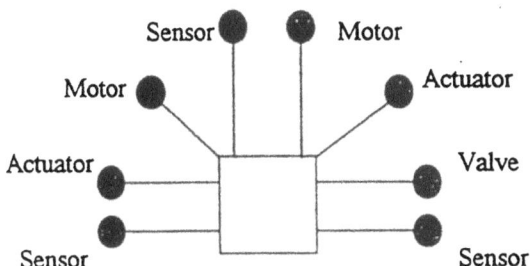

Figure 1 : Typical Centralised Control System.

In this kind of configuration, when the central microcontroller is not functioning, the whole system is also malfunctioning. Also, the number of IO expansions is limited by the number of IO ports available in the microcontroller. Hence, IO expansion is not easy and wiring cost increases with the number of devices added to the system.

The distributed control system eradicates all these disadvantages by connecting all the sensors and devices to a local operation network (LON). LON is a technology that allows devices like actuators and sensors to communicate with

one another through an assortment of communication media using standard protocol.

A distributed control system (DCS) is one in which all the nodes are networked together in a medium using twisted pair wires. The nodes communicate in this pair of wires by means of a common protocol.

One of the advantages of this method is that node expansion is easy. A node can be added to the network by just connecting it to the network wires. Wiring is also minimised especially if the number of nodes is large. Also, single point of failure will not halt the whole system.

Some of the designs and evaluation of DCS can be found in [4]. A typical DCS system in which all the nodes are connected in a bus topology is as shown in figure 2.

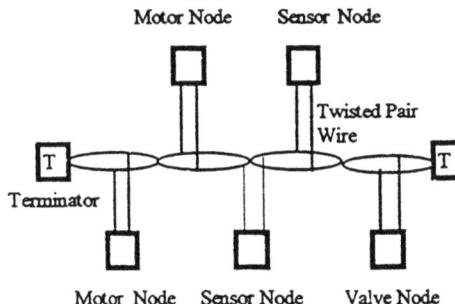

Figure 2 : Typical Distributed Control System

2   HOME AUTOMATION MODEL

Numerous literatures have reported projects on home automation [5]. However, home automation using DCS technology is still quite conservative. Our laboratory developed such model, where all the nodes are connected together in a bus topology as shown in figure 3.

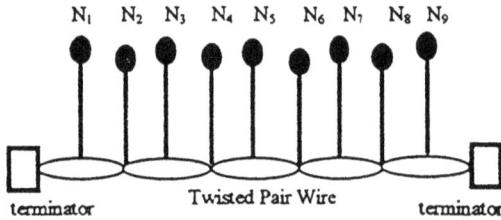

Figure 3 : Home Automation Model

The model consists of a temperature sensor & cooler node (N_1), brightness sensor node(N_2), curtain motor control node(N_3), lighting control node(N_4), infra-red receiver node(N_5), smoke-detector node(N_6), motion-detector node(N_7), security alarm panel node(N_8), door-contact & release node(N_9).

The brightness sensor node senses the outdoor brightness and feedbacks this to the curtain motor which opens and closes the curtain accordingly. The temperature sensor node feedbacks the room temperature to the cooler which cools the room accordingly. The infra-red node activates the dimmers in the lighting nodes. The smoke detector node activates the security panel alarm when smoke is detected. The door contact/release node gives door automation capabilities that can detect forced entry, door is unlocked etc. The motion detector node detects intrusion and sends an alarm to the security.

All the network variables of the model communicate using the LONWORKS technology which is a control network technology developed by Echelon Corporation for monitoring and controlling devices. This technology makes use of a chip called the Neuron Chip which can communicate with one another in a peer-to-peer manner using the LONWORKS protocol. The nodes are interconnected using a pair of twisted pair wire which allows network variables to be communicated between the nodes. The media

access control (MAC) uses the predictive p-persistent CSMA/CD protocol.

3   NEURON 3150 CHIP

The model was developed based on the Neuron 3150 chips from Echelon Corporation. The architecture of the chip [1] is as shown in figure 4.

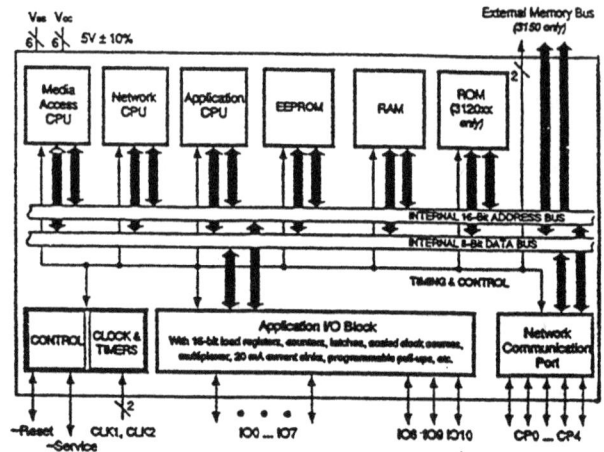

Figure 4 : Neuron Chip Block Diagram

The Neuron chip mainly consists of the on-chip EEPROM, the RAM, the ROM, the 11 bi-directional I/O pins for interfacing to the application hardwares, the network transceiver interface, timer/counter hardwares, the Media Access Control CPU, the Network CPU and the Application CPU.

The microcontroller chip has three CPUs which are the application CPU, the Network CPU and the Medium Access Control (MAC) CPU. The Media Access Control (MAC) CPU handles layers one and two of the seven-layer OSI framework. The first layer is the physical layer which defines the methods used to transmit and receive the data on the network. The second layer is the data-link layer which performs framing, data encoding, error checking, collision avoidance and detection.The application CPU handles the user written application codes. The network CPU which implements layers 3 and 4 takes care of the networking management, network variable processing, addressing, authentication etc.

The I/O pins can be configured in different ways for different applications. The different types of I/O objects configurations that can be supported at the present moment are summarised in figure 5, details of which are available in [1],[3].

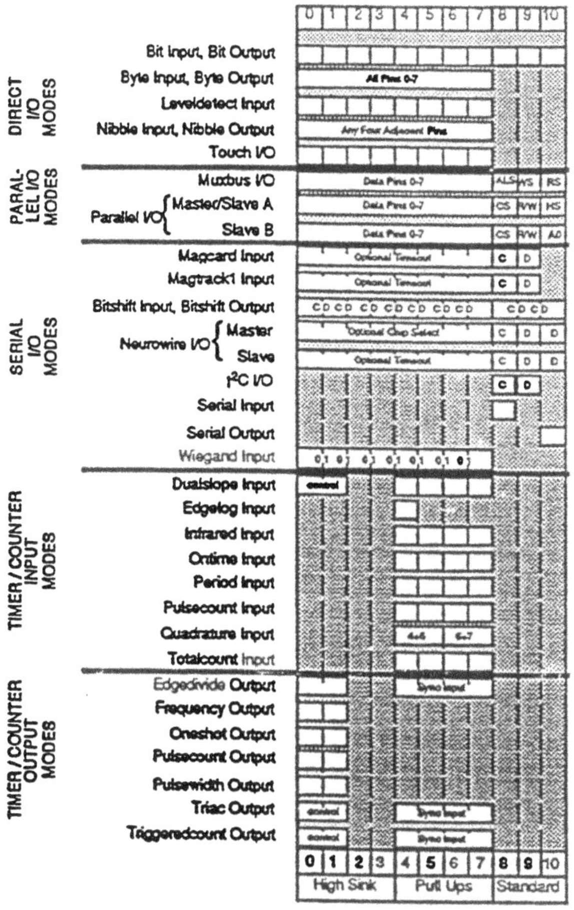

Figure 5: I/O Object Pin Configuration Summary

4 NETWORK MEDIA ACCESS PROTOCOL

The system uses the LonTalk predictive p-persistent Carrier Sensed Multiple Access with Collision Detection (CSMA/CD) MAC network protocol. The p-persistent CSMA/CD rules are:

(1) If the medium is idle, transmit with probability p; otherwise probability (1-p). delay one time slot.

(2) If the medium is busy, continue to listen until the channel is sensed idle; then repeat step 1.

(3) If a collision is detected during transmission, transmit a brief jamming signal to assure that all nodes know that there has been a collision and cease transmission.

(4) After transmitting the jamming signal, wait a random amount of time, then attempt to transmit again. (go to step 1).

The LonTalk protocol used by the chip has an added improvement over the p-persistent CSMA/CD protocol. It ensures that p is dynamically changed based on the network load. When the network is idle, all nodes randomise over only 16 slots whereas for a heavy load, the number of slot increases. This will prevent channel idleness on light load and low collision rate on heavy load. The estimation is based on the fact that the transmitting node carries information in the packet on the number of acknowledgements expected as a result of sending that packet. The details on the protocol can be obtained in [2].

5 NETWORK VARIABLES

The nodes talk to one another by means of network variable propagations in the twisted pair wires. The data items in the presentation layer are called the network variables which can be single data item or data structure.

In order for any node to understand the network variables from another node, the network variables need to be connected. Network variable connection or binding is a process whereby nodes' network variable inputs are linked to nodes' network variable outputs. The binding process causes the generation of address information which is used by the nodes to route information from one node to another. For example, considering two nodes A and B having the following programs:

Node A's Program
//Push Button IO
IO_4 input bit io_pb;

//network variable nvo_x declaration
network output boolean nvo_x;

```
//Push button pressed task
when(io_changes(io_pb) to 1)
  {
  nvo_x = 1;// Propagate nvo_x as 1
  }
```

Node B's Program
```
//LED IO
IO_0 output bit io_led;

// declare nvi_x as input network variable.
network input boolean nvi_x;

// task when there is an update in nvi_x
when(nv_update_occurs(nvi_x))
    {
    static boolean status;
    status = 1-status;
    // toggle on/off the LED
    // when nvi_x is updated.
     io_out(io_led, status);
    }
```

In order for "nvi_x" to receive information from "nvo_x", the two variables must be connected by the network binder. When the variables are connected, the LED of node B will toggle on/off each time a "nvi_x" is updated.

To expand the LED node further, we can add another LED node (node C), which is exactly similar to node B, to the network and bind up the network variables. The network nodes are then reloaded with the new network binding information and the system still works.

6   CONTROL NETWORK INSTALLATION

Connecting all the nodes to the twisted pair wires is insufficient for the system to work. Another step which is the installation process is required to be performed. In this process, the nodes are given the network configuration information, such as the network addresses, to enable them to talk to the right node. The network configuration information is defined and loaded with an installation tool provided by the developement tool.

There are three basic steps to installation. Firstly, each node is given a logical address assignment. This address uniquely identifies the node in the network. The second step is the binding step which involves connecting up the network variables together by means of the netowrk variable binder. The last step is the configuration which is the downloading of the configuration information into the nodes. More details of the control network installation overviews can be found in [6].

7   CONCLUSIONS

A home automation model using DCS technology was developed using the Neuron chip 3150. Future projects will involve free topology instead of the bus topology. The free topology will be implemented by changing the control modules to the free topology modules. The rest of the circuit remains intact.

ACKNOWLEDGEMENTS

The authors wish to acknowledge the support provided by Ngee Ann Polytechnic Electrical Engineering Department. The authors would also like to thank the HOD/Mr Goh, Dr Schafer, Mr Herz, Mr Glatter and the students for their participation.

REFERENCES

[1]   Echelon Corporation (1995), Neuron Chip Data Book,Palo Alto, USA.

[2]   Echelon Corporation (1995) ,Enhanced Media Access Control with Echelon's Lontalk Protocol. *LONWORKS Engineering Bulletins.*

[3]   Motorola (1997), *LonWorks Technology Device Data Book.*

[4]   Michael P. Lukeas (1986), *Distributed Control Systems, Their Evaluation and Design*, NY Van Nostrand Reinhold Co.

[5]   Delban T. Horn (1991), *Home Remote Control and Automation Project*, TAB Book.

[6]   Echelon Corporation (1995), LONWORKS Installation Overview, *LONWORKS Engineering Bulletins.*

DISTRIBUTED IMPLEMENTATION OF COMMUNICATING PROCESS ARCHITECTURES FOR AUTONOMOUS MOBILE ROBOTS

Gen'ichi Yasuda

*Department of Mechanical Engineering, Nagasaki Institute of Applied Science,
536 Aba-machi, Nagasaki 851-01, Japan*

Abstract: This paper describes the implementation of a parallel distributed control architecture for autonomous mobile robots. Based on a modular and hierarchical approach, the control architecture is defined as a set of communicating sequential processes of autonomous sensing and control modules. The distributed implementation of the process interaction network is realized through synchronous and asynchronous communications among concurrent processes. A method for coordinating obstacle avoidance and trajectory tracking using fuzzy logic is presented based on the concurrent sensing and control framework. The control architecture is successfully implemented for local control and servo control tasks using a network of microcontrollers connected via a serial bus.

Keywords: Communicating process architectures; multiprocessing systems; software development; distributed control; autonomous mobile robots.

1. INTRODUCTION

Autonomous mobile robots should be able to perform multiple tasks by coordinating distributed sensors and actuators in a coherent way. The control system of mobile robots should include adaptive behaviour to react with noisy sensors and imperfect actuators under uncertain or unexpected situations. As the capability of a mobile robot increases, a number of sensors and actuators must be incorporated and coordinated with sophisticated computing components (Hu and Brady, 1996). So, it is very important to design the architecture for the control system of autonomous mobile robots.

A behaviour based control architecture has been proposed for robot navigation in unknown environments (Brooks, 1986). Robot tasks are decomposed into several simple reactive or sensorimotor behaviours, on the basis of the stimulus-response scheme in biological organisms,

and performed without an entire world model and complex reasoning processes. However, in complex environments, lower-level behaviour control should be fused with higher-level global planning and control for the efficient coordination of conflicts and competitions among multiple reactive behaviours.

This paper describes a parallel distributed control architecture which takes the form of a network of communicating processes of autonomous sensing and control modules. Each autonomous module has its own knowledge-based coordination process, communication links, and facilities to access external devices. Based on a concurrent sensing and control framework, a method of sensor-based local control for obstacle avoidance using fuzzy logic is presented. Using low-cost, efficient microcontrollers, the distributed implementation of a motion control system for power wheeled steering type mobile robots is also described to show the effectiveness of the proposed control architecture.

2.DISTRIBUTED CONTROL ARCHITECTURE

The distinguishing features of control systems for autonomous mobile robots are described, and a distributed control architecture is defined as a set of communicating sequential processes of autonomous sensing and control modules.

2.1 Description of the Control Architecture

Control systems for autonomous mobile robots can be hierarchically composed of several different levels of control. Generally, these include task planning, global control, local control, and servo control levels. The hierarchy of control corresponds to a hierarchy of information resolution (Isik and Meystel, 1988). During the course of the robot motion, as the updated downward and upward information flow continues, a refined trajectory control is brought about along the approximately planned path. The control of the robot motion at lower levels is performed through more detailed information obtained from multi-resolutional external and internal sensors at various proximities. The conceptual hierarchical structure of the control system is shown in Fig.1.

Sensing, perception and control are distributed among different levels for global control, obstacle detection, local control, and steering and speed control tasks. Each control level constitutes its own control loop. Different control layers need different responce times. At a level, when behaviour is suppressed owing to any disturbance or any abnormal signal is received, then control at the next upper level is invoked. Further, controls at any upper levels can be invoked. The coordination process at a level can send a command to the coordination process at any lower level or generate commands employing behaviours at all lower levels as well as at the same level.

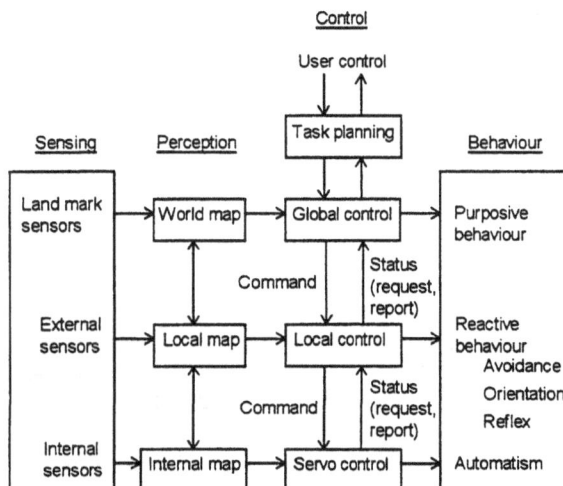

The entire system can be built up stepwise from low-level behaviours like automatism, reflex behaviour, and obstacle avoidance to high-level purposive behaviours like path planning, by analogy with biological organisms. Constraints and commands with parameters are passed downwards, and different status are passed upwards. The control system will suffice with just the final point to successfuly reach the destination point.

The functions of each control level are the following. A mobile robot is operated by the user's command with imprecise descriptive information about the robot motion and its environment. The task planning level, supervised by the user, determines the tasks to be done in terms of physical objects in the environment. The global control level, based on incomplete or topological world description, approximately provides a set of control points or subgoals through which the robot should to try to pass by. These control points are expressed in terms of desired position, orientation, speed, and a type of precision of passing or docking points.

The local control level repeatedly takes a subgoal from the global control level and calculates the relative direction and distance of the new coordinates of the subgoal. Then, considering obstacle avoidance based on sensory information about the uncertain environment, it issues steering and speed orders to the servo control level. The servo control level offers a virtual mobile base accepting orders regarding desired steering and speed. For power wheeled steering type mobile robots, the servo control level evaluates the reference values of axial movement of two independently drived wheels, then controls the angular velocity of each.

In trajectory control, the local control level tries to reach each subgoal fed by the global control level as desired. When any obstacle is detected by an external sensor, avoidance control is brought about. As a result, when a subgoal is surpassed, unreachable or missed, it reports the status to the global control level and requests for a new one. The global control level may recalculate the successive subgoals.

2.2 Communicating Process Network

The proposed distributed control architecture is composed of a number of concurrent sensing and control processes and expressed as a set of communicating sequential processes (CSP) (Hoare, 1985). The control architecture can be implemented using a general parallel processing programming language, such as occam and the Inmos parallel C language derived from CSP.

Fig.1. Hierarchical structure of the control system.

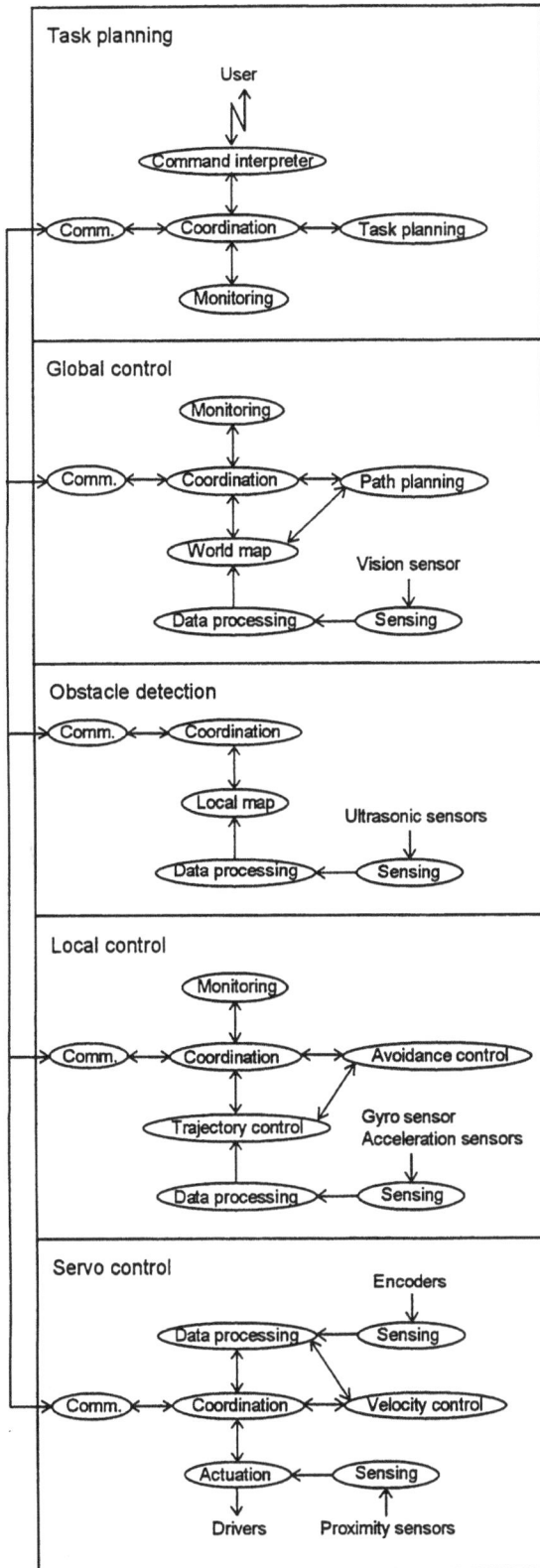

Fig.2. Communicating process network for autonomous mobile robots.

The communicating process network is shown in Fig.2. The configurations of processes in all the control levels are similar and modularized for nested hierarchical control. Each coordination process is equipped with a knowledge-based controller and performs the synchronization of activities of the

sensing and control processes in the same level. In particular, at the local control level, a fuzzy logic based navigation controller coordinates lower-level obstacle avoidance and high-level global path planning. Each of the two axial servo control processes, based on the orders issued by the local control process, moves the robot to the desired steering angle and speed.

All levels in the control system are intended to operate in parallel to achieve concurrent actions. Interactions in the communicating process network are implemented using synchronous and asynchronous communications (Yasuda and Tachibana, 1997). Within each control level synchronous communications are used, because the specific behaviour at the level should be synchronized with the corresponding sensory signals. On the other hand, between different levels, asynchronous communications are used. Synchronous and asynchronous interaction control processes are used for distributed implementation of the communicating process network. I/O-parallel blocks are used in order to avoid deadlock in a process network composed of a large number of processes (Welsh, 1987).

The control flow structure of the discrete event coordination and sequencing executed on the servo control process is modelled using a Petri net as shown in Fig.3. In servo control of two independently driven wheels, mutual communication between servo control processes is required for synchronization of servo control cycle. The servo control process continues until the specified subgoal is reached, determined unreachable or a new command is given by the coordination process of the local control process. The Petri net models incorporate well known Petri net interpretations of occam constructs, such as ALT constructs (alternation), WHILE constructs (conditional loop), and IF constructs, and include explicit models of the interprocess communications.

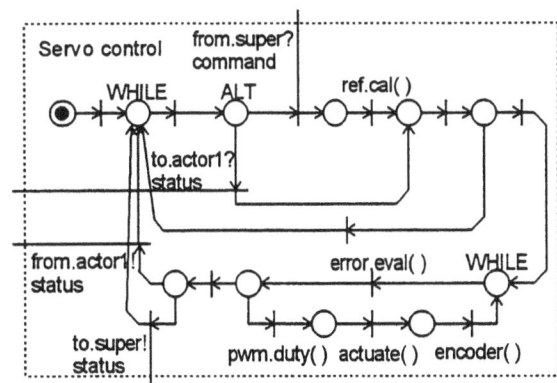

Fig.3. Petri net representation of servo control process for angular velocity control.

3. SENSOR BASED LOCAL BEHAVIOUR CONTROL USING FUZZY LOGIC

Sensor based behaviour control using fuzzy sets and fuzzy reasoning may improve navigation performance, because mobile robots must be controlled using noisy sensors and imperfect actuators in dynamic environments subject to unpredictable disturbances. The fuzzy rules for navigation are naturally decomposed into several sets of rules, each set deals with a sensing or control subsystem for navigation. This decomposition method can be implemented as real-time software, since the number of fuzzy rules in each set is rather small. Thus, the fuzzy rules can be easily constructed for more complex environments. The building block method can be used for realizing the implementation of sets of rules on a parallel distributed control architecture. The computational structure using fuzzy logic for obstacle avoidance and target steering is composed of three concurrent processes as shown in Fig.4.

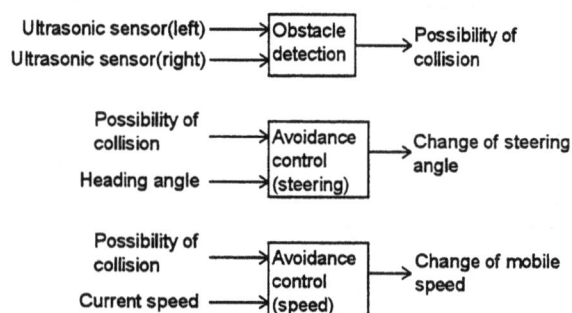

Fig.4. Fuzzy reasoning processes for obstacle avoidance.

The input signals to the obstacle detection process are the distances between the robot and obstacles to the left and right locations. An ultrasonic sensor composed of one transmitter and two receivers mounted on the top of the mobile robot is used to detect the distances at the right and the left in front of the robot in terms of the time of flight. When the obstacle is a wall, the inclination of its surface can be inferred or estimated. To cope with inaccuracies and uncertainties of sensor data, the sensor readings of the two receivers are fuzzified, using the three linguistic fuzzy sets: *far*, *medium*, and *near*. Then, the results are integrated to infer the possibilities of collision at the right, left and center in front of the robot using fuzzy reasoning (Vazquez and Garcia, 1994). Linguistic fuzzy set values and their meanings for the possibilities of collision are as follows:

RO: *right out of center*   RC: *right center*
LC: *left center*   LO: *left out of center*
LW: *low possibility*

Then, fuzzy reasoning of the change in steering angle and the change in speed for avoidance control are separately performed. The change of steering angle is inferred in consideration of the possibilities of collision and the direction of a specified subgoal. The direction is defined as the heading angle between the robot and the subgoal. All directions around the robot are divided into eight sections to fuzzify the heading angle. Linguistic fuzzy set values and their meanings are as follows:

FF: *front*   BB: *back*
FL: *front and left*   BR: *back and right*
LL: *left*   RR: *right*
BL: *back and left*   FR: *front and right*

Linguistic fuzzy set values and their meanings for the change of steering angle are as follows:

LS: *left small*   RS: *right small*
LB: *left big*   RB: *right big*
ZZ: *zero*

A set of fuzzy rules for the change of steering angle is shown in Table 1. When the robot is moving to a specified subgoal through a narrow channel, or escaping from a U-shaped obstacle, it must reflect following edge behaviour. The fuzzy rules show the robot follows an edge of an obstacle when the obstacle is close to the right (left) of the robot, and the subgoal also is located to the right (left). When the acquired information from the ultrasonic sensor shows that there are no obstacles around the robot, its main reactive behaviour is steering for the subgoal, and the fuzzy rules show that the robot quickly adjusts the heading angle to the subgoal.

Table 1 Fuzzy rules for change of steering angle

	FF	FL	LL	BL	BB	BR	RR	FR
RO	LS	LS	LB	LB	LB	LB	LS	LS
RC	LS	LB	LB	LB	LB	LB	LB	LB
LC	RS	RB	RB	RB	RB	RB	RB	RB
LO	RS	RS	RS	RB	RB	RB	RB	RS
LW	ZZ	LS	LB	LB	LB	RB	RB	RS

Change in mobile speed is inferred in consideration of the possibilities of collision and the current speed. When information acquired by the ultrasonic sensor shows that there exist obstacles nearby the robot, or the robot moves at curved and narrow areas, its speed is reduced to avoid obstacles. The current speed of the robot is fuzzified using the five linguistic fuzzy set values: *very slow, slow, medium, fast,* and *very fast*. The speed change of the robot is fuzzified using the five linguistic fuzzy sets: *decrease, slight decrease, no change, slight increase,* and *increase*.

Each membership function in the sets of fuzzy rules is represented by a triangle, while both sides of outer region are exceptionally defined as a combination of ramp and step functions. By the min-max inference algorithm and the centroid defuzzification method, fired rules are weighted to determine an appropriate control action.

By integration of different reactive behaviours, such as obstacle avoidance and decelerating at curved and narrow areas, following edges, and steering for the subgoal, rather than by firing a single behaviour, a specified subgoal can be reached. The coordination can be easily performed using fuzzy reasoning. The weights of the obstacle avoidance and target steering behaviours depend largely on the possibilities of collision; while the weight of the following edge behaviour depends on the heading angle.

The fuzzy rules were constructed based on the human reasoning and tuned by iterative simulations. It is rather easy to construct the fuzzy rules and not time-consuming to tune the constructed rules appropriately. The effectiveness of the proposed avoidance method were verified by a series of simulations.

4. IMPLEMENTATION ON A NETWORK OF MICROCONTROLLERS

The proposed control architecture takes the form of a network of sensing and control modules and can be implemented in many different ways independent of any particular processor, although Inmos transputers seem to be the most appropriate as the basic processor. To examine the effectiveness of the proposed control architecture, a power wheeled steering type mobile robot was developed. As the first step to parallel distributed implementation, the PIC16C74 microcontrollers (Microchip Technology Inc.) were used as building blocks to implement the distributed sensing and control architecture. They employ an advanced RISC architecture with separate instruction and data buses. Because of the two stage instruction pipeline, all instructions are efficiently executed in a single cycle, except for program branches. For real-time control applications, several peripheral features are available including: three timer/counters, two PWM modules, two serial ports, and an 8-channel high-speed 8-bit A/D converter which is suited for applications requiring low-cost analog interface.

A synchronous serial port can be configured as the two-wire Inter-Integrated Circuit (I^2C) bus developed by the Philips Corporation. The I^2C interface employs a comprehensive protocol to ensure reliable transmission and reception of data. When transmitting data, one device is the "master" and generates the clock, while the other devices act as the "slave". All portions of the slave protocol are implemented in the synchronous serial port module's hardware, while portions of the master protocol need to be addressed in the PIC software. The protocol sequence for initiating and terminating data transfer allows a master to send "commands" to the slave and then receive the requested information or to address a different slave device.

The output stages of the clock and data lines must have an open-drain in order to perform the wired-AND function of the bus. When two or more masters try to transfer data at the same time, arbitration on the data line and clock synchronization occur using a wired-AND connection. The interrupt generation on start and stop bits in hardware facilitates software implementation of the master functions.

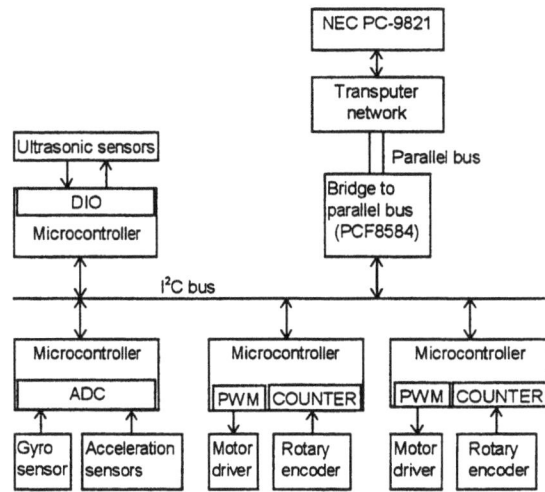

Fig.5. Hardware configuration of the control system.

The hardware configuration of the constructed control system is shown in Fig.5. The four microcontrollers are connected via an I^2C bus to perform the lower-level control. A transputer network or other general-purpose high-performance processors for numerical computation, which perform the higher-level control, are connected with the microcontrollers through a bridge to a parallel bus. The control system realizes the local control and servo control levels. Three microcontrollers perform servo control tasks, while one microcontroller is dedicated to obstacle detection. The transputer network is in charge of the local control.

The local control computer executes obstacle avoidance and trajectory control. Preliminary experimental results show that fuzzy logic based navigation control enables the robot to avoid obstacles and reach a specified subgoal in a more stable manner than crisp control based on a subsumption framework, where a single behaviour is fired at a time and it is difficult to adjust their thresholds and other field parameters.

In the servo control microcontrollers, for trajectory tracking, the orientation and the position of the robot are measured by using a piezoelectric vibratory gyroscope and two variable capacitance type acceleration sensors in the same way as ordinary inertial navigation systems. An angular velocity signal from the gyro sensor is sampled at intervals of 10 ms, converted through an A/D converter, and integrated to obtain the orientation of the robot. The measurements of output voltage of the gyro sensor and the acceleration sensors are confirmed to be reliably in proportion to the angular velocity and the acceleration, respectively. Each acceleration signal from the sensors are sampled at the same intervals and integrated to obtain the two-dimensional velocity and displacement of the robot.

The computational structure for trajectory tracking executed in servo control is shown in Fig.6. The position and orientation errors are used to measure the trajectory tracking performance. The angular velocity difference between the left and right wheels can be set equally from the desired forward angular velocity and determined as a function of the position and orientation errors. The linear feedback control law can be replaced by a fuzzy logic control law, where the position and orientation errors are input variables and the angular velocity difference is the output variable and these variables are fuzzified to form fuzzy values by using appropriate membership functions. The application of a tuned fuzzy logic control law shows an improved trajectory tracking performance than crisp control laws.

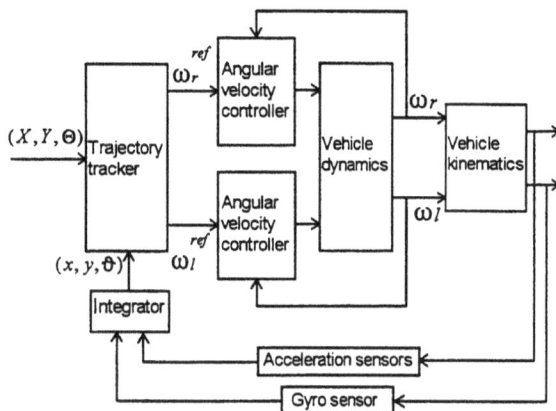

Fig.6. Feedback control structure for trajectory tracking.

Macro codes for 16-bit and 32-bit arithmetic computation in PID servo control are written in the assembly language. The cycle time of 50 ms is possible at the local control level including fuzzy

reasoning for obstacle avoidance; while the cycle time of 10 ms is possible at the servo control level including the localization using the gyro sensor and acceleration sensors. For four microcontrollers and one bridge connected via a shared bus, bus contention does not bring about serious limitations. The capability of parallelism also improves the response time. These satisfies the requirements for the real-time constraints imposed by the lower levels of the control system. When more efficient processors, such as Inmos transputers, are used, the performance can be largely improved, because of their fast interprocessor communication speed and microcoded multitasking scheduler.

5. CONCLUSIONS

A modular and hierarchical approach to building a sensor-based distributed control architecture for mobile robots has been described. As the first step to parallel distributed implementation, low-cost one-chip microcontrollers were used as building blocks to implement the distributed sensing and control architecture. The constructed control system performs concurrent sensing and actions for obstacle avoidance and trajectory tracking, and reduces the difficulties of complex software development.

REFERENCES

Brooks, R.A. (1986). A robust layered control system for a mobile robot. *IEEE J. of Robotics and Automation*, **RA-2**, 14-23.

Hoare, C.A.R. (1985). *Communicating Sequential Processes*. Prentice-Hall International, London.

Hu, H. and M. Brady (1996). A parallel processing architecture for sensor-based control of intelligent mobile robots. *Robotics and Autonomous Systems*, **17**, 235-257.

Isik, C. and A.M. Meystel (1988). Pilot level of a hierarchical controller for an unmanned mobile robot. *IEEE Trans.*, **RA-4**, 241-255.

Vazquez, F. and E. Garcia (1994). A local guidance method for low-cost mobile robots using fuzzy logic. *Proc. IFAC Symposium on Artificial Intelligence in Real Time Control*, 203-207.

Welsh, P.H. (1987). Emulating digital logic using transputer networks (very high parallelism = simplicity = performance). *Lecture Notes in Computer Science*, **258**, 357-373.

Yasuda, G. and K. Tachibana (1997). Implementation of communicating sequential processes for distributed robot system architectures. *Preprints of the IFAC-Workshop on Manufacturing Systems: Modelling, Management and Control*, 349-354.

EVALUATION OF NETWORK PROTOCOL
FOR AUTOMOTIVE DATA COMMUNICATION

Jung-A Yun*, Sang-Woon Nam*, and Suk Lee**

* : *Graduate School, Pusan National University*
** : *School of Mechanical Engineering, Pusan National University*

Abstract : The growing number of electronic components used in automobiles has
given rise to problems concerning the increasing number, size and weight of the
wiring harnesses. As one approach to resolve these problems, multiplexed wiring
systems using automotive communication protocol such as Controller Area
Network(CAN), Advanced PALMNET and J1850 have been developed by many
automobile companies. In order to compare these protocols quantitatively, this paper
presents the performance evaluation of CAN and Advanced PALMNET via discrete
event simulation. Through numerous simulation experiments, several important
quantitative performance factors such as the probability of a transmission failure,
average system delay(data latency), and throughput have been evaluated.

Keywords : automotive control, communication protocol, discrete-event systems,
distributed computer control systems, network, performance evaluation, simulation
language, time delay

1. INTRODUCTION

The major part of advances in automobile
technology has been propelled by the dramatic
progress in electronics. In addition to the
conventional switches, sensors, and electric loads,
today's car has various electronic components in
order to enhance the automobile performance
including safety, emission control, and
convenience by using Anti-lock Braking
System(ABS), Engine Management System(EMS),
and so on. These various components are
spread throughout a vehicle, and are connected
to each other by separate wiring.
As the number of electronic components
increases, the following problems have been
identified :
- excessive wiring weight that adversely
  affects fuel economy
- lack of installation space
- increased cost due to numerous wires and
  connectors
- inflexibility with respect to design changes
  assembly
- difficult diagnostics of electrical system

In order to overcome the above technical
difficulties, multiplexed wiring technology has
been adopted as a new approach to connecting
those electronic components. The key concept
of the multiplexed wiring is digital serial
communication through a shared transmission
medium similar to a local area network. This
concept has its root in the fly-by-wire technique
for military and commercial aircraft that connects
various aircraft components by the same
principle. With the multiplexed wiring, a number
of network interface modules or stations are
installed at various locations collecting signals
from not only control-related components such
as ECU's (Electronic Control Units) and various
sensors but also switches, lamps, and so on. In
this way, a great number of wires and
connectors can be replaced with a single network
cable, which reduces wiring weight, requires only
the minimal installation space, and makes the
assembly process much easier. Furthermore,
location of a component has less meaning in
terms of wiring, which makes the design change
and diagnostics easier.
In order to achieve these benefits several
automotive network protocols, which is

mentioned above have been already developed. For instance, there are Controller Area Network (CAN) by Bosch (Philips Semiconductors, 1991; Robert Bosch GmbH), Advanced Protocol for Automotive Low and Medium speed Network (PALMNET) by Mazda (Kimura, et al., 1994; Inoue, et al., 1989), J1850 by Society of Automobile Engineers (Society of Automobile Engineers, 1995), and so on. As the domestic companies are trying to employ this technology, it is necessary to select a proper protocol throughout the objective evaluation among many network protocols. In order to estimate the system delay (time interval between the message generation and the completion of its transmission) and throughput (transmission capability of useful information in a unit time) which is the heart of the network performance, discrete event simulation (Law, et al., 1990) is used. The evaluation via simulation has its advantage since this method can be executed under realistic situations without assumptions for the analytical technique.

In this paper, we present the simulation models for CAN and Advanced PALMNET protocol by a discrete event simulation language called SIMAN (Pegden, et al., 1990) and the evaluation results of CAN and Advanced PALMNET. This paper consists of five sections including introduction. Section II describes the working principles of CAN and Advanced PALMNET while section III explains the developed simulation models. The results of the simulation is presented in section IV. Finally, section V presents the conclusions.

2. CAN AND ADVANCED PALMNET

CAN, a serial communication protocol, was developed by Bosch for connecting controllers, sensors, and actuators in automobile. CAN employs a type of distributed access control (Philips Semiconductors, 1991; Robert Bosch GmbH) that provides the equal control rights for every station. The characteristics of CAN include high transmission rate, flexibility, high data integrity, prioritized access control, data consistency, and error detection function.

On the other hand, PALMNET was developed as LAN system of the distributed control system in a vehicle, and it was successfully installed in Eunos Cosmo by Mazda (Kimura, et al., 1994). On the basis of this PALMNET protocol, Advanced PALMNET is developed (Inoue, et al., 1989). Advanced PALMNET uses the distributed access control like CAN, and offers the data rate ranging from 125Kbps (kilobits per second) to 1Mbps (megabits per second).

2.1. Medium Access Control

Medium access control is the function determining which station should transmit the message because many stations share a common network. CAN and Advanced PALMNET use the same distributed access control. An ID (identifier) is given to each message in order to identify the message and to control the bus access. Being different from the general LAN protocol, CAN and Advanced PALMNET assign ID according to the contents of a message (content-based addressing). For example, the message for transmitting engine rpm (revolution per minute) has the same ID regardless of generating station. In this content-based addressing, a station wishing to transmit sends data and ID together, and other stations receive or ignore the message depending on the ID.

This assigned ID classifies the message and is used to define the priority of each message. That is, when more than one stations attempt to transmit concurrently, stations may have to stop its transmission after comparing its ID value with others. Figure 1 shows the block diagram of the operational logic of a station.

Each station constantly monitors the bus to determine whether any station is using the network. Because the electric signal on the network has a finite speed of propagation, there is always a chance for multiple stations to start their transmission within a short time interval. This situation, called message collision, is resolved by comparing the identifiers of the involved messages.

That is, the message with the lowest identifier value wins the right to use the network while others stop their transmission immediately.

Since this identifier-based bitwise arbitration does not lose any information and time, it is called non-destructive bus access. Unlike CSMA/CD (Carrier Sense Multiple Access with Collision Detection) which is a well-known example of distributed bus access, the identifier-based bitwise arbitration allows the network to be utilized efficiently even under heavy network load.

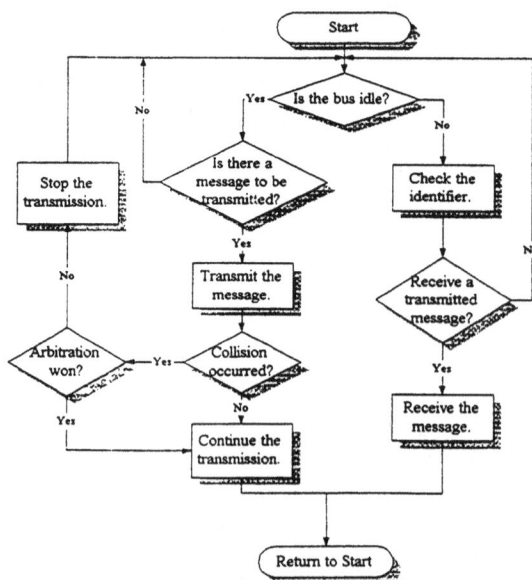

Fig. 1. Operational logic of a station

2.2. Protocol Layer for Automobile

CAN supports the data rate ranging from 5Kbps to 1Mbps, with either linear bus or star topology. Protocol structure of CAN (Robert Bosch GmbH) is made up of three layers : physical layer, transfer layer, and object layer. The physical layer defines signal level, bit representation method, transmission medium and so on. The transfer layer transfers the received messages to the object layer and accepts messages from the object layer. More specifically, the transfer layer has the function of fault confinement, message collision arbitration, and error detection. Also, the transfer layer turns the messages into the frame and sends the acknowledgement when messages are transmitted properly. The object layer provides an interface to the applications.

Layer structure of Advanced PALMNET (Inoue, et al., 1989) is divided into a physical layer and a data link layer according to the OSI(Open Systems Interconnection) reference model (Stallings, 1993). The physical layer defines the transfer medium, bus interface, bit representation method, transfer rate, and so on. The data link layer defines access form, priority control, error detection, acknowledgement, etc.

2.3. Message Format

CAN defines four types of message packets which is called frame : data frame, remote frame, error frame, and overload frame. A data frame is used to send up to 8 bytes data to other devices and consists of seven fields as shown in Table 1. In the table, each bit can take two values : 'd'(dominant, numeric value of 0) and 'r'(recessive, numeric value of 1).A remote frame has the same structure of the data frame except the data field and it is used to request the transmission of a data frame. Being different from a data frame, the remote frame has RTR bit value 'r', and a data length code (DLC) has to be identical to the one included in the data frame. In addition, an error frame is used to notify all devices on the network of an error while an overload frame is transmitted when a device is not ready to receive another frame. All frames are separated from the previous frame by the interframe space.

Data frame format of Advanced PALMNET is explained in Table 2.Start of Frame(SOF) is consisted of 8 bits, PRI(frame priority) is used to control the bus access and the priority. In contrast to CAN, there is an independent ID field to indicate the contents of a frame. Also the TYPE field indicates the type of a frame, e. g., SDG frame and data frame. Data field can use up to 4 bytes for medium or low speed and uses 4 bytes or 8 bytes for high speed according to whether the frame is for the SDG mode which enables one station to gather data from multiple stations simultaneously.

Table 1. Data frame format of CAN

BIT VALUE	FIELD NAME AND FUNCTION	FIELD LENGTH
d	**Start of Frame** · notify start of message and synchronize all stations	1
	Arbitration Field · Identifier :11bits	12
d	· RTR Bit : 1 bit	
d d	**Control Field** · Reserved bits : 2 bits	6
	· Data Length Code(DLC) : define data length of data in data field (4 bits wide)	
	Data Field · use the message of data length 0 for synchronizing distributed process	0 ℓ 64
	CRC Field · CRC Sequence : be made after transferring the last bit of data field. check the error by using the shift register of 15 bits	16
r	· CRC Delimiter : 1 bit	
r	**ACK Field** · be consisted of ACK slot and ACK Delimiter.	2
r	· after receiving message correctly the station transfers 'd' bit to the transmit station and ACK slot exchanges 'r' bit for 'd' bit.	
r .... r	**End of Frame** · notify the end of message.	7

Table 2. Data frame format of Advanced PALMNET

BIT VALUE	FIELD NAME AND FUNCTION	FIELD LENGTH
d .... d r d	**Start of Frame(SOF)** · notify start of message and synchronize all stations	8
	Frame Priority(PRI) · define the priority.	8
	Frame Type · define the form of transfers frame	8
	Data Identifier(ID)	8
	Data Field · middle or low speed : 4bytes · high speed : for SDG mode 4 bytes or 8 bytes	32 or 64
	CRC Field · code for error detection · middle or low speed : 8 bits · high speed : 16 bits	8 or 16
	End of Data(EOD) · only use on high speed · be used to switching the bit synchronization	2
	ANC(Acknowledgement for Network Control) Field · transfer node of the message sends the acknowledgement. · middle or low speed : 17 bits(including ACK conviction bit) · high speed : 16 bits	16 or 17
	End of Frame · notify the end of message.	10

For error detection, CRC (Cyclic Redundancy Check) field is used. If a station receives data correctly, then ANC (Acknowledgement for Network Control) bit is set. The ANC field collects the acknowledgement signals from multiple stations in one transmission frame. Finally, EOF notifies the end of a frame and let the waiting stations begin to transmit.

3. THE DEVELOPMENT OF SIMULATION MODEL

For a message ready to be transmitted, the time interval needed for its successful transmission is a random variable. Some messages must wait until the on-going transmission is finished or, due to collision with other messages, requires several retransmission. This kind of transmission delay is affected by several factors that include the number of stations connected to the network, the frequency of message generation, and the length of message. The messages which is transmitted and received in a vehicle system, need to be transmitted within a limit of time because they contain some information needed to operate a vehicle, e.g., sensor signal and control command. For a network system design, it is necessary to have some preliminary knowledge about time delay until the successful transmission. In order to obtain this kind of quantitative performance, simulation of the network is needed and a simulation language called SIMAN is used to develope the simulation models of CAN and Advanced PALMNET.

The developed simulation model accepts as its input the number of stations, transmission speed, and the statistical information about the time distribution of message generation and the length of messages. The outputs include average or standard deviation of system delay, the number of messages transmitted, the frequency of message collision, the throughput and the rejected number of the messages.

This simulation model assumes the normal operation of the network. Individual stations belong to one of four traffic groups with different probability distributions of the message generation time and the message length. While the both protocols compare ID field and PRI field bit by bit in case of collision, the simulation model compares them with the decimal values at the end of the ID or PRI field.

Both simulation models have common structure and consist of an initialization subroutine called PRIME, four events, and two subroutines. The instant of execution and the functions of each event and subroutine are explained in table 3. The simulation model can be represented by figure 2 where the solid arrow represents scheduling of the next event with a non-zero delay, the dashed arrow represents scheduling without time delay, and a node represents an event or a subroutine.

Table 3. Function of simulation model subroutines

subroutine	function	timing
PRIME	* the initialization of parameters * scheduling the first messages arriving at each stations	the beginning of simulation
MSG_AR (EVENT 1)	* scheduling next arrival of messages in each stations * giving attribution to the messages arriving * inserting the arriving messages into the queue * calling TRANS if there is message to transmit without any other transmission message	the arrival of message
TRANS	* scheduling the end of transmission message * scheduling the arrival of head of message in each station	
END_TX (EVENT 2)	* When transmission ends normally, - removing the oldest message in the queue - scheduling the tail of the message arriving in each station - calling TRANS if any messages in the que	the end of transmitting message
MSG_HD (EVENT 3)	* When the message lost in comparison of ID - canceling the schedule of END_TX - calling CANCEL, subroutine - immediately executing END_TX	
MSG_TL (EVENT 4)	* When there are messages in the que and no stations transmit - calling TRANS	receiving the last bit of message
CANCEL	* canceling the scheduled events	

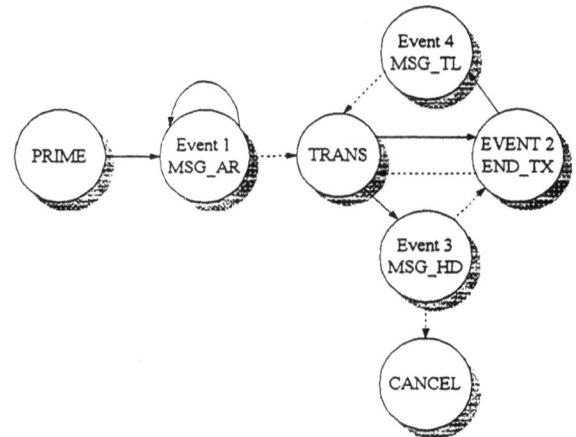

Fig. 2. Event diagram for simulation model

4. PERFORMANCE EVALUATION

The CAN and Advanced PALMNET networks has been simulated with the following assumptions.

- The messages generated at all stations have the identical statistical characteristics.
- The generated messages are data frames of constant field length.
- The time interval between two adjacent message generation at a station is uniformly distributes over 512 μs to 1536 μs.

With these assumptions, the number of stations are varied from 2 to 16, and the length of data field, is increased from 0 to 8 bytes, while the network is operating with the transmission speed of 1Mbps. The table 4 is showing the data traffic that is the ratio of the data transmission load (except overhead) to the total capacity of the network. The probability of transmission failure due to the change of the data field and the number of stations is shown in figure 3 and figure 4 for CAN and Advanced PALMNET, respectively. The average system delay appears in figures 5 and 6, and the throughput is shown in figures 7 and 8.

Table 4. Data traffic according to simulation conditions

		Data Field (byte)								
		0	1	2	3	4	5	6	7	8
Number of stations	2	0	0.016	0.031	0.047	0.062	0.078	0.094	0.109	0.125
	4	0	0.031	0.062	0.094	0.125	0.156	0.188	0.219	0.250
	6	0	0.047	0.094	0.141	0.188	0.234	0.281	0.328	0.375
	8	0	0.062	0.125	0.188	0.250	0.312	0.375	0.438	0.500
	10	0	0.078	0.156	0.234	0.312	0.391	0.469	0.547	0.625
	12	0	0.094	0.188	0.281	0.375	0.469	0.562	0.656	0.750
	14	0	0.109	0.219	0.328	0.438	0.547	0.656	0.766	0.875
	16	0	0.125	0.250	0.375	0.500	0.625	0.750	0.875	1.000
Simulation time		time for transmitting the messages of number 1500								
Queue capacity		1								
Message arrival period	distribution	Uniform								
	mean	1024 [512 ~ 1536] (μs)								

The probability of transmission failure is defined as the number of transmission failure divided by the number of transmission attempts. CAN and Advanced PALMNET have the similar tendency as shown in figure 3 and 4. Namely, as the data field length and the number of stations increases, the probability of transmission failure increases. As the data field gets longer, there will be more chance of other messages waiting during a message transmission. Also, as the number of stations increases, there will be more chance of waiting message during a fixed length of a message transmission. The only difference between two protocols appears when there are many stations with short data fields. For CAN, the probability of transmission failure remains small while that for Advanced PALMNET increases rapidly as the number of stations increases. This difference comes from the fact that Advanced PALMNET has much larger overhead compared to CAN.

Even with no data byte, Advanced PALMNET's frames are long enough to use substantial portion of the network capacity, which results in higher probability of transmission failure.

Figures 5 and 6 show the average system delay. System delay is defined as the time interval from the instant when a message is generated to the instant when the message's last bit is transmitted. Like the probability of transmission failure, it can be seen that the average system delay increases as the data field length and the number of stations increase.

This phenomenon is due to the fact that the higher traffic causes the longer transmission time, which implies that other frames have to wait longer.

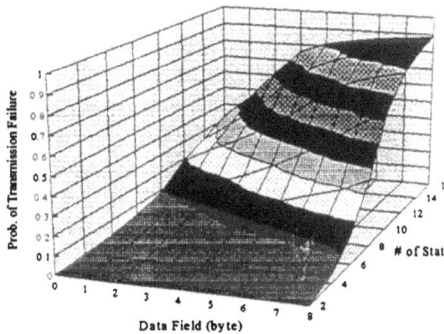

Fig. 3. Probability of transmission failure of CAN

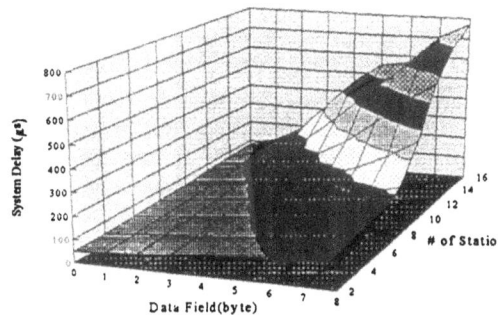

Fig. 5. Average system delay of CAN

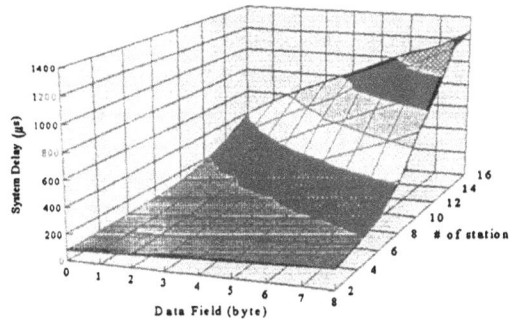

Fig 4. Probability of transmission failure of Advanced PALMNET

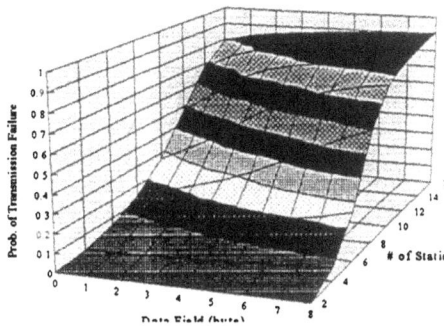

Fig. 6. Average system delay of Advanced PALMNET

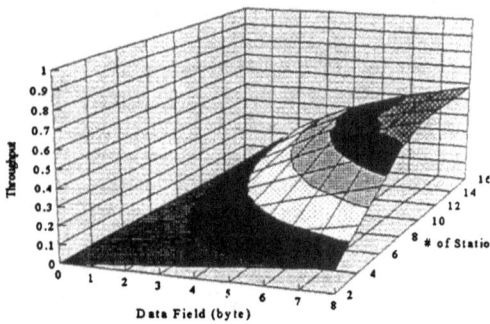

Fig. 7. Throughput of CAN

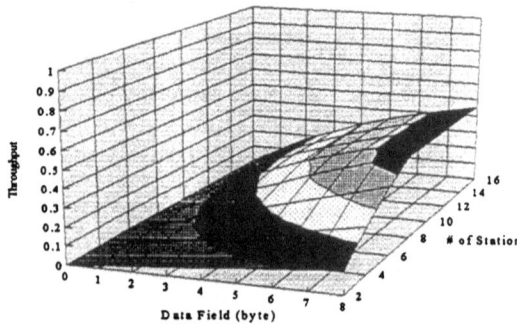

Fig. 8 Throughput of Advanced PALMNET

Also, the system delay increases because of the increase in the probability of transmission failure. It can be also observed that long overhead of Advanced PALMNET leads to relatively larger system delay than CAN.

Figures 7 and 8 are illustrate the trend of throughput, i.e., the ratio of the amount of data transmission to the total available transmission capacity of the network. As the data field length and the number of stations increase, throughput increases up to a certain point.

As traffic increases, throughput begins to level off and converges eventually to a certain limit. Since most of frames are transmitted in case of low traffic, the throughput has the same trend as traffic. But the higher traffic causes the increase in collision probability and system delay. This results in increase in the probability that the newly generated frame is rejected because the transmission queue is still occupied by the previous frame.

Therefore, the traffic continues to increase but the throughput remains at the same value. Even though the trend of throughput is almost identical for both CAN and Advanced PALMNET, CAN shows the higher throughput.

5. CONCLUSION

In this research two simulation models have been developed for Advanced PALMNET (Protocol for Automotive Local Area Multiplexing Networking) and CAN (Controller Ares Network). These simulation models can be an essential tool to design a vehicle network by estimating quantitative performance factors such as system delay and throughput. Using these simulation models, the performance of CAN and Advanced PALMNET has been evaluated and the conclusions are listed below.

1) Both protocols exhibit the similar characteristics in terms of the probability of transmission failure, average system delay, and throughput. As traffic increases, average system delay increases due to the increased chance of collisions and the increased time of transmission.

2) CAN appears to have better performance compared to Advanced PALMNET mainly because of its short overhead. However, this conclusion is made without considering efficient utilization of the network capacity by use of SDG and ANC features of Advanced PALMNET.

3) Since the worst-case average system delay is around 1 ms, both protocols seem to be suitable for distributed real-time control with the sampling time of tens of ms. However, the network designer should avoid higher traffic because important frames can be lost due to rejection of a newly generated message.

REFERENCES

Hideki Kimura, Yusaku Himono et al. (1994). The Development of the Advanced Protocol for Automotive Local Area Multiplexing Network (Advanced PALMNET), *SAE paper,* **940365.**

Law, A. M. and W. D. Kelton. (1990). *Simulation Modeling and Analysis,* 2nd ed. McGraw-Hill.

Pegden, C., R. Shannon and Sadowski. (1990). *Introduction to Simulation Using SIMAN,* McGraw-Hill.

Philips Semiconductors. (1991). *CAN Specification Version 2.0*

Ray, A. (1987). Performance Evaluation of Medium Access Control Protocols for Distributed Digital Avionics, *Journal of Dynamic Systems, Measurement and Control,* **Vol.109.**

Robert Bosch GmbH. *Controller Area Network (CAN) -The Breakthrough for the Networking of Electronics in the Vehicle*

Society of Automobile Engineers. (1995). *SAE J1850*

Stallings, W. (1993). *Local and Metropolitan Area Networks,* Macmillan

T.Inoue et al. (1989). Protocol for Automotive Local Area Network (PALMNET)- A Newly Developed In-Vehicle Communication System Based on SAE J1850, *SAE Paper,* **890535**

Upender., B. P. (1994). Analyzing the Real-time Characteristics of Class C Communications in CAN Through Discrete Event Simulations, *SAE paper,* **940133.**

IMPLEMENTATION AND PERFORMANCE EVALUATION OF PROFIBUS IN THE DISTRIBUTED COMPUTER CONTROL SYSTEMS

Seung Ho Hong[*], Ki Am Kim[**]

[*]*Department of Control and Instrumentation Eng., Hanyang University, Korea*
[**]*Hyudai Information Technology Co., Ltd., Korea*

Abstract: This paper presents an implementation method of Profibus interface S/W based on the FMS(Fieldbus Message Specification) and FMA7(Fieldbus Management Layer 7). The Profibus interface S/W is implemented on PC and embedded controller. In order to enable the Profibus interface S/W handle many application tasks and communication services, two kinds of real-time/multi-tasking operating system, CTask and OS-9, are utilized. Using the Profibus interface S/W, this study develops an experimental model of DCCS(Distributed Computer Control Systems), and evaluates the message delay in the user layer. Message delay is measured with respect to the changes of message length, message generation time and TRT(Token Rotation Time). The outcomes of experiment are compared with those of the simulation model which comprises only the physical and data link layers of Profibus. This study investigates the relationship between the message delay in the data link layer and the message delay up to the user layer.

Keywords: Profibus, FMS, FMA7, interface S/W, experiment, performance

1. INTRODUCTION

Large and complex systems such as automated factories and DCCS(Distributed Computer Control Systems) in the steel, chemical and power plants comprise several distributed and spatially dispersed field devices. Interconnection of these field devices through networking increases reliability, flexibility, and user friendliness in design, operation and management of complex systems. Fieldbus is the lowest level industrial network in the computer communication hierarchy of factory automation and DCCS. Fieldbus provides digital, bi-directional, multidrop, and serial-bus communications among isolated field devices such as supervisory computers, controllers, PLCs, transducers, actuators, sensors and operator station(Jordan, 1995; InTech, 1996; Piementel, 1990).

This paper presents an implementation method of fieldbus interface S/W. The fieldbus interface S/W is developed on the basis of FMS(Fieldbus Message Specification) which is adopted as the

application layer protocol in the Profibus(DIN 19 245, 1991; Bender, 1993) and Foundation Fieldbus(Fieldbus Foundation, 1996). In this study, the Profibus interface S/W is implemented on PC and IUC(Intelligent Universal Controller). Industrial PCs are widely used as general controllers, and data processing, logging and monitoring units in the DCCS(Babb,1996), and IUC is a modular-type commercial embedded controller(PEP Modular Computers, 1993).

Fieldbus interface S/W usually handles many application tasks and communication services. In this study, CTask(Wagner, 1990) and OS-9 (Dibble,1994) are used as real-time/multi-tasking operating system in PC and IUC, respectively. Using the Profibus interface S/W, this study develops an experimental model of DCCS, and evaluates the performance of Profibus network. The experiment measures user level message delay with respect to the changes of message length, message generation time and TRT(Token Rotation Time). The outcomes of experiment are compared with those of the simulation experiment which models only the physical and data link layers of Profibus. From the results, this study investigates the relationship between the message delay in the data link layer and the

This work was supported by KOSEF under Grant 961-0924-138-2.

message delay up to the user layer. Based on the experimental results, this study suggests some factors that should be considered in the design and implementation phases of the Profibus-based DCCS.

This paper consists of four chapters. Chapter 2 describes the structure of the Profibus interface S/W developed in this study. Chapter 3 presents the experimental model, and analyzes the results of the experiment. Finally, chapter 4 presents the conclusions and future works of this study.

2. DEVELOPMENT OF PROFIBUS INTERFACE S/W

This chapter describes the structure of the Profibus interface S/W developed in this study. The Profibus interface S/W utilizes FMS(Fieldbus Message Specification) and FMA7(Fieldbus Management Layer 7) services provided by the application layer of Profibus. Section 1 of this chapter, describes the implementation of Profibus interface S/W on PC. Profibus interface S/W on IUC is discussed in section 2.

2.1 Implementation of Profibus Interface S/W on PC

Industrial PCs are widely used in many application fields of DCCS. This section presents an implementation method of Profibus interface S/W on PC. PC is connected to Profibus network using a Profibus controller board. The essential function blocks of a Profibus controller board include serial interface to Profibus, central control unit, time management and data interface to application systems. In this study, CP5412 board(Softing, 1993) is used as a Profibus controller board.

The fieldbus interface S/W usually handles many application tasks and communication services simultaneously. In this paper, application task implies the application program executed in the DCCS, and communication service implies the exchange of data and information between application programs and Profibus application layer. The real-time data generated from field devices must be completely transmitted within a specified bound of time interval. Thus, it is desirable to develop a fieldbus interface S/W in real-time/multi-tasking programming environment. In this study, PC-based Profibus interface S/W is developed using CTask which provides real-time/multi-tasking kernels on the DOS. CTask is a public software and its functions are limited. However, the programming skill presented in this study can be applicable to any higher level commercial software.

In this study, Profibus interface S/W is implemented such that application program can transmit and receive data using the services provided by FMS and FMA7. In the following, the FMS and FMA7 services provided by Softing (Softing, 1994) is briefly described. Fig. 1 shows the structure of data exchange between

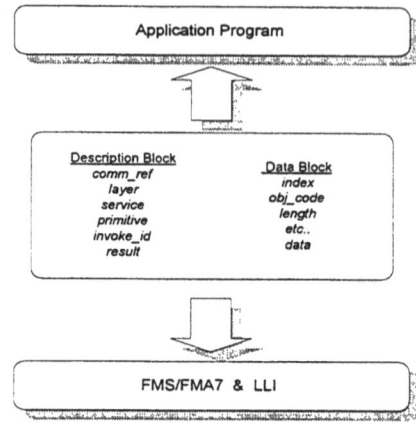

Fig. 1. Data exchange between FMS and application program

application programs and FMS/FMA7. As shown in Fig. 1, the exchange is carried out through description block and data block. Description block describes the communication service that is currently executed, where *comm_ref* refers the communication relationship which is already defined in CRL(Communication Relationship List). *layer* distinguishes FMS and FMA7, and *service* specifies the current service provided by FMS or FMA7. FMS is operated by client-server relationship, and the service is executes through *primitive*. *invoke_id* is the identifier of invoked task and *result* indicates the result of the specified service. Data block indicates the actual data that is transmitted or received. It refers OD(Object Directory) using *index*, *obj_code*, etc. The information related to the description and data blocks must be pre-registered in CRL and OD during the initialization stage of network system. LLI(Lower Layer Interface) generates data frame using description and data blocks, and transfers it to the FDL(Fieldbus Data Link) layer(DIN 19 245, 1991).

Fig. 2 shows the structure of the interface S/W which is developed using CTask. The tasks of *receive_cnf_ind*, *transmit_req_res* and FMS are concurrently executed, and the application programs can be added as much as they needed. When a new data is received, *receive_cnf_ind* assigns an index number of the data, and saves the corresponding description and data blocks. The index number is also saved in the index_queue so that it can be referred by FMS.

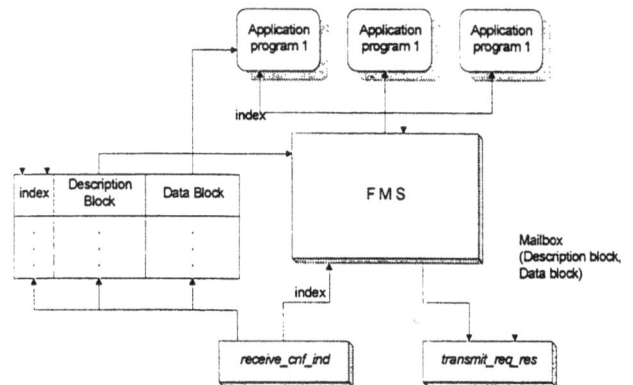

Fig. 2. Structure of PC-based Profibus interface S/W

If FMS detects a new index, it examines the description block of the received data, and transfers the index number to the corresponding application_index_queue so that it can be referred by the application program. If application program detects a new index while executing its own task, it accepts the receiving data by examining the data block, and continues to execute its own task. If application program produces a new data to transmit, it generates the description and data block of the corresponding data and sent it to *transmit_req_res* using mailbox. *transmit_req_res* checks the mailbox to find a new data to be transmitted. If a new data is detected, *transmit_req_res* generates PDU (Protocol Data Unit) using the description and data blocks of the corresponding data, and transfers it to the lower layer. In this Profibus interface S/W, the application programs can execute their own tasks while transmitting and receiving the data through Profibus.

Fig. 3 to 5 show the task sequences of *receive_cnf_ind*, *transmit_req_res* and FMS, respectively. As shown in Fig. 3, *receive_cnf_ind* initializes the queue and index when the task starts. When a receipt of new data is detected by a confirm or indicate primitive from FDL (Fieldbus Data Link) layer, *receive_cnf_ind* assigns an index number of the data, and saves the corresponding description and data blocks in the queue, and also saves the index number in the index_queue so that it can be referred by FMS.

Fig. 4 shows the task sequence of *transmit_req_res*. *transmit_req_res* checks the mailbox to find a new data to be transmitted. If a new data is detected, *transmit_req_res* generates service primitive using the description and data blocks of the corresponding data, and transfers it to FDL

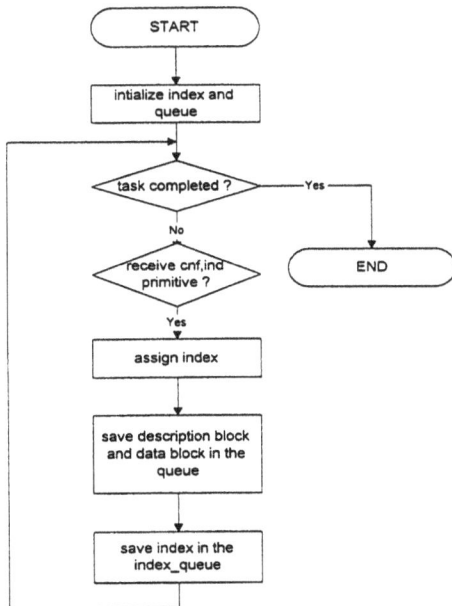

Fig. 4. Task sequence of *transmit_req_res*

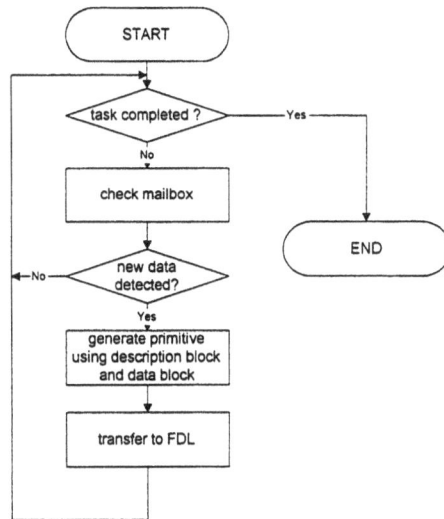

Fig. 5 Task sequence of *FMS*

Fig. 5 shows the task sequence of FMS. If FMS task detects a new index in the index_queue, it examines the description block of the received data. If the received data requests rejection of connection or event management service, FMS task executes error or event management service. If not, FMS task transfers the index number to the corresponding *application_index_queue* so that it can be referred by the application program.

Fig. 6 shows a simple example of application program. If application program detects a new index in the *application_index_queue* while executing its own task, it examines the data block of received data, updates the corresponding data in the application program, and continues to execute its own task. If application program produces a new data to transmit, it generates the description and data block of the corresponding data and sent it to *transmit_req_res* using mailbox.

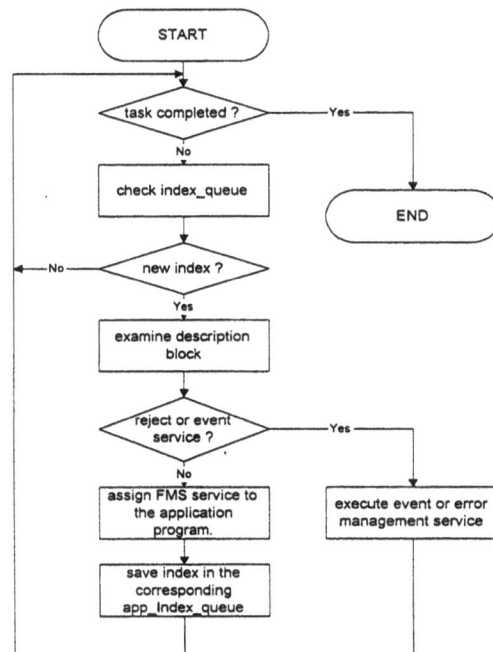

Fig. 3. Task sequence of *receive_cnf_ind*

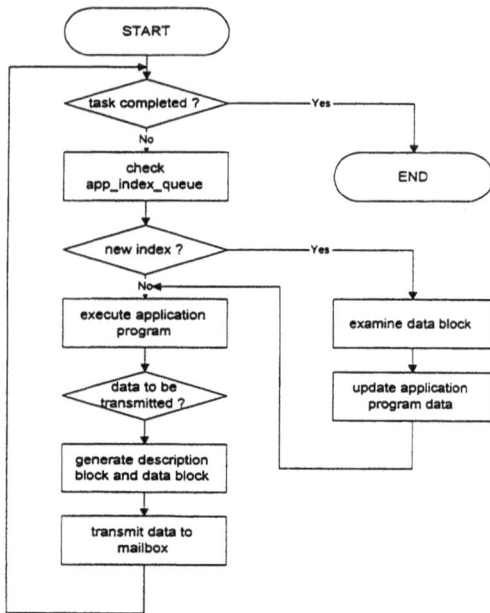

Fig. 6. Task sequence of application program

2.2 Implementation of Profibus Interface S/W on IUC

DCCS frequently use embedded controllers for some specific tasks. This section describes the method of implementing Profibus interface S/W in the embedded controller. In this study, IUC is used as an embedded controller. IUC is a modular-type embedded controller which has built-in Profibus controller. In this study, Profibus interface S/W for IUC is developed using the real-time/multi-tasking operating system of OS-9.

Fig. 7 shows the structure of Profibus interface S/W. The structure is very similar to the case of PC-based interface S/W. Data communication among the tasks are executes through pipe. It also has three basic tasks of receive_cnf_ind, transmit_req_res and FMS, and their task sequences are identical to those of the PC-based interface S/W

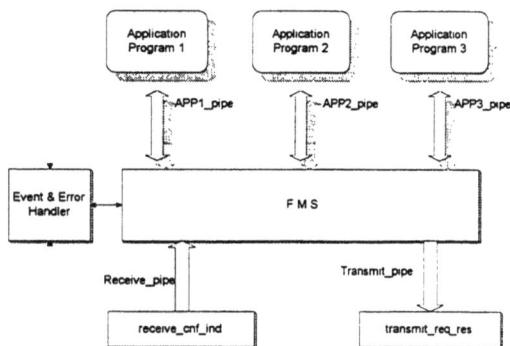

Fig. 7. Structure of Profibus interface S/W developed using OS-9

3. MESSAGE DELAY IN THE EXPERIMENTAL MODEL

Based on the Profibus interface S/W described in

the previous chapter, this study develops an experimental model of automatin system. The experimental model consists of 1 PC, 3 IUCs and 1 Smart I/O. Smart I/O is a kind of IUC which has built-in functions of I/O module and PLC. The automation systems such as robot controllers and machine controllers can be interconnected through PC and IUCs, and the field devices such as sensors and actuators can be interconnected through Smart I/O. Experimental model also comprises one bus monitor in order to examine the status of the data frame. Fig. 8 shows the configuration of experimental model.

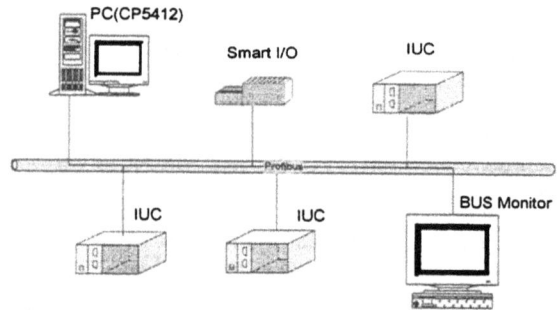

Fig. 8. Experimental model of DCCS

The major factors which can affect the performance of Profibus network include message interarrival time, message length and TRT(Hong, 1994). Using the experimental model, this study measures the user level message delay with respect to the changes of message interarrival time, message length and TRT. As shown in Fig. 9, the message delay is measured from the instant when PC requests write service to one of the IUCs to the instant when its response is arrived at the PC. During this transaction, other nodes generate predefined amount of network traffic. In order to clarify the analysis of experimental results, message interarrival time, message length and TRT in all nodes are set to identical.

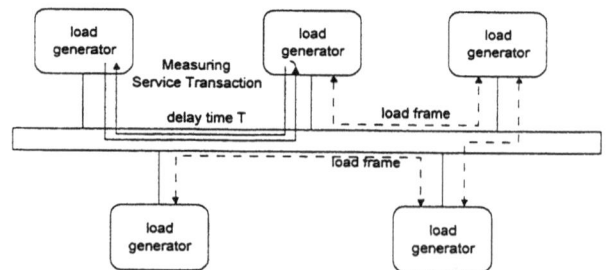

Fig. 9. Measurement scheme of message delay time

Message delay is measured with respect to the change of message interarrival time, message length and TRT. The results of experimental model are compared with those of the simulation model which comprises only the physical and data link layers of Profibus(Hong, 1994). This study investigates the relationship between the message delay in the data link layer and the message delay up to the user layer level.

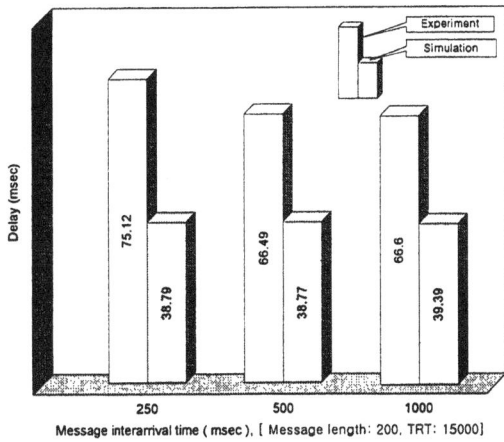

Fig. 10. Delay with respect to the change of message interarrival time

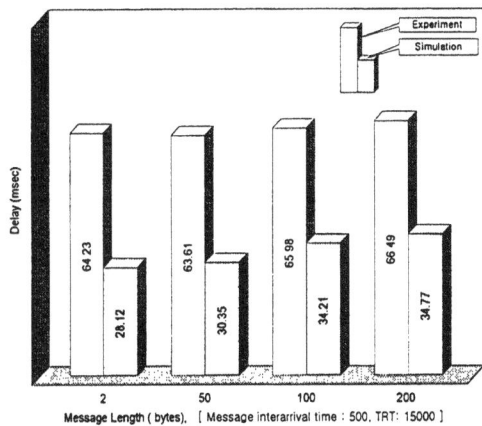

Fig. 11. Delay with respect to the change of message length

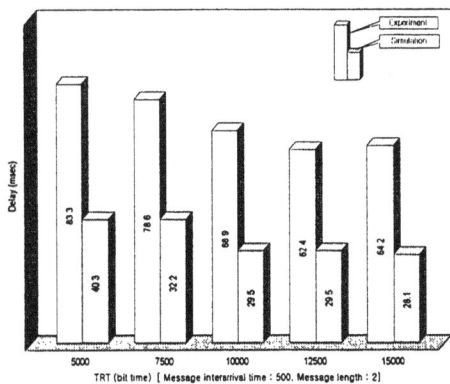

Fig. 12. Delay with respect to the change of TRT

Fig. 10 shows the comparison of the user level message delay and the data latency up to the data link layer when the message length generated from all nodes in the network is 200 bytes and TRT is set to 15000 bit time. The change of message delay in the data link layer(simulation results) is not so significant because the offered traffic in the network is quite low. However, even in low traffic, message delay up to the user layer is considerably affected by the change of message interarrival

time because the processing time in the application program of the user layer is increased as the number of message is increased.

Fig. 11 shows the effect of the change of message length on the message delay when all the nodes in the network generate their data with an interval of 500 msec and TRT is set to 15000 bit time. Fig. 11 shows that the delays in both user and data link layers are increased as the message length is increased. The rate of increment at the data link layer, however, is higher than that at the user layer level. Fig. 11 shows that, even though the change of message length affects more sensitively on the delay of data link layer, its effect on the data latency up to the user layer is not so significant. This is because the processing time at the user layer is not uniform.

Fig. 12 shows the effect of TRT on the message delay when all the nodes in the network generate their data with an interval of 500 msec and the message length is 2 bytes. All the nodes in the network generate low priority message only. As expected, message delay is increased as TRT is decreased because the low priority message can be transmitted only when the actual token rotation time is smaller than the setting value of TRT. Fig. 12 shows that the message delay in both user and data link layers are quite sensitively affected by the change of TRT.

The experimental and simulation results suggest that the following factors should be considered when a Profibus-based DCCS is designed and implemented.

(1) In general, message delay up to the user layer is quite larger than that up to the data link layer. User layer message delay can be increased especially when messages are frequently generated in the application programs. Thus, it is not suitable to design a network system only based on the results of simulation model which usually models up to the data link layer of network system. Network system designer should consider the safety factor that compensates the additional delay occurred in the application and user layers.

(2) Compared to the changes of message interarrival time and message length, the change of TRT shows significant influence on the delay performance in both user and data link layers. The change of TRT plays a significant role even up to the user level data latency. Thus, the TRT-based priority mechanism needs to be appropriately utilized when real-time and non-real-time data traffic are co-exist.

4. CONCLUSIONS AND FUTURE WORKS

This paper presents an implementation method of Profibus interface S/W which can be applicable to the real-time process of DCCS. The Profibus interface S/W is developed using the real-time/multi-tasking environments of CTask and OS-9. This study also develops an experimental model of DCCS using the Profibus interface S/W, and measures message delay in the user layer. The experimental results are compared with the results obtained from the simulation model that comprises physical and data link layers. This study investigates the relationship between the message delay up to the data link layer and the message delay up to the user layer level.

This study is an on-going work. Currently, an experimental model of automation system is developed. The experimental model consists of two robots, two conveyer belts, one NC machine, several sensors and an operator station. These automation devices exchange their data using the Profibus interface S/W developed in this study. The experimental model will be actively utilized in studying appropriate method of implementing Profibus network in the manufacturing automation systems.

REFERENCES

Babb, M. (1996, January). "PCs: The Foundation of Open Architecture Control Systems" *Control Engineering*.

Bender, K. (1993). *PROFIBUS-The Fieldbus for Industrial Automation*, Carl Hanser Verlag & Prentice Hall

Dibble, P. (1994). *An Advanced Programmers Guide to OS-9*, Microware Systems.

DIN 19 245, (1991). *Profibus Standard Part 1*

DIN 19 245, (1991). *Profibus Standard Part 2*

Fieldbus Foundation, (1996). *Foundation^{TM} Specification: Fieldbus Message Specification.*.

Hong, Seung Ho. (1994, April). "Approximate Analysis of Timer-Controlled Priority Scheme in the Single-Service Token-Passing Systems,"*IEEE/ACM Trans. on Networking*, Vol. 2, pp. 206-215.

InTech, (1996, November). *Field Buses Special Issue*, ISA Publication.

Jordan, J. R. (1995). *Serial Networked Field Instrumentation*, John Willey & Sons.

PEP Modular Computers, (1993). *IUC User's Manual*.

Piementel, J. R. (1990). *Communication Networks for Manufacturing*, Prentice Hall.

Softing GmbH, (1993) *Hardware description PROFI-IF-PCAT, ISP-IF-PCAT*.

Softing GmbH, (1994) *Profibus Communication interface Layer7 for CP5412-A1 Controller*.

Wagner, T. (1990). *CTask: A Multitasking Kernel for C*, Ferrari Electronic GmbH.

PRE-RUN-TIME SCHEDULING METHODS OF DATA LINK LAYER IN THE FIELDBUS ENVIRONMENT

Young Shin Kim and Wook Hyun Kwon

Control Information Systems Lab., School of Electrical Eng.
Seoul National University, Seoul, 151-742, Korea
Phone: +82-2-873-2279, Fax: +82-2-878-8933
E-mail: kys@isltg.snu.ac.kr

Abstract: This paper deals with real-time properties, scheduling of data link layer and time and space consistencies of application processes in FIP. In doing so, the paper recommends the values of data link layer timers and shows the transmission performance of FIP. And the paper suggests two kinds of scheduling methods of data link layer in the fieldbus environment to satisfy the scheduling constraints such as memory size, medium utilization, and communication jitters. In addition a method of guaranteeing time and space consistencies of application processes is also proposed.

Keywords: Fieldbus, FIP, Real-time, Pre-run-time scheduling, Memory size, Network utilization, Communication jitters

1. INTRODUCTION

As distributed control system spreads wide in the industrial environment, the need for real time properties of network increases. Fieldbus has been suggested to satisfy such needs and also introduced to replace the traditional wiring of sensors and actuators with a simple network. Fieldbuses are networks which have been and are being used in the various industrial fields, and therefore they have experienced various changes and have been improved according to the characteristics of the fields.

Since most time-critical communications are likely to occur between two devices in the same work-cell, a multi-access bus is a natural candidate for connecting devices in a workcell. Mini-MAP, CSMA/CD, Profibus, FIP, etc. are of such topology. Mini-MAP has been introduced to reduce the communication delay, but it still leaves the real-time issue unaddressed. CSMA/CD type protocols can not be applicable to real-time systems because of their unbounded communication delays. Profibus has almost the same architecture as

mini-MAP so it has the same problem. But FIP supports real time communication by reserving the required communication capacity in advance for each real-time channel, and by using a centralized scheme for scheduling information in order to guarantee the delivery of real-time information.

FIP, however, often face with not a few problems when it is applied to the real fields. Efficient management of communication media and support of real-time properties are important ones of them. First, to use communication media efficiently in using fieldbuses can be regarded as to send data to the destination correctly in time, to receive data from the source correctly in time and to support predictability for the whole system by guaranteeing response time for a request. This problem, in other words, is how to schedule communication media in applying fieldbuses to the industrial fields. Scheduling problem often occurs in various fields of application whenever it is necessary to assign a common resource to one of several processes requesting it, at the same time respecting the processes' time constraints. Respecting time constraint takes on various forms according to the context in which scheduling is being considered. Pre-run-time scheduling is essential when timing constraints should be applied to hard-real-time systems (Jia Xu, 1991). Some results on pre-run-time scheduling are to reduce the scheduling size (S. Cavalieri, 1995) in fieldbus environment, and to make scheduling distance uniformly in application processes (Ching-Chih Han, 1992).

Second, in most industrial communication networks physical layer and data link layer support real-time properties to some degree. But real-time properties of application layer or application process levels have not been addressed because it is strongly related with software architecture and scheduling of application processes, so it is hard to deal with real time concept in those areas. Furthermore this problem should substantially touch the synchronization mechanism between hardware and software. So, the problem has been studied only by a few researchers though it plays an important role in many applications. However, most of these efforts (P. Lorenz, 1995; P. Pleinevaux, 1988) only touched the real-time issues in fieldbuses. At the application processes, however, real-time communication between devices in an automated factory has seldom been made. This means that though the first problem of how to use communication media efficiently is resolved there is still a very big obstacle to be resolved as before. In other words, even if we use a very good scheduling method and get a very good schedule data and then apply the schedule to the fieldbus network, we can obtain only the data link layer performance not application layer or application process level performace, which means the network does not support application process level real-time. If we use only data link layer we will meet more complicated problems of how to dispense with the various function and facilities of application layer. In conclusion it is necessary to support application process level real-time and show the real-time bound to support the system predictability.

In Section 2, some real time issues of FIP are discussed. In Section 3, several scheduling methods are applied to the scheduling of data link layer and a guideline for choosing one of them to satisfy scheduling constraints is suggested. In Section 4, a method of guaranteeing time and space consistencies of application processes is suggested. Conclusion is in Section 5.

2. REAL TIME ISSUES

FIP is divided into three layers whose respective functions correspond to those defined in the OSI model. They are physical layer, data link layer and application layer. Each layer has its own unique role and cooperates with other layers to support good communication services. FIP has several kinds of real-time issues such as Data Link layer timer setting, transmission performance, scheduling of Data Link layer, Time and Space consistencies of application processes and so on.

2.1 Data Link Layer Timer

FIP has seven timers in data link layer (DLL) which are shown in Table 1 (WorldFIP, 1995). They support timing functionality of DLL. Data link timer setting is about timing of physical layer. It determines such as when to send next signal and how long to give synchronization padding signal and so on. Therefore it mainly depends on the characteristics of the physical medium and once the values of the timers are set at the configuration stage there is almost no possibility to be changed.

Table 1. Data link layer timers

Timer	Location	Function
T0		Max. silent time
T1	BA	Max. absent time of RPxx after IDxx
T2	BA	Length of synchronization window
T3	BA	Max absent time of IDxx after RPxx
T4	CS	Max. absent time of RP_DATxx after ID_DAT
T5	BA	Max. absent time of RP_END after ID_MSG
T6	MSS	Max. absent time of RP_ACK after RP_MSG_ACK

BA: Bus Arbitrator, CS: Consuming Station, MSS: Message Source Station

Some recommendation values of timers are as follows. A reference timer T0 is defined as corresponding to the maximum silent time permitted on a network segment. The timer T0 is a global systems parameter. Five timers are defined using T0 as a basis. The T1, T4 and T6 timers correspond to the use of T0 in the specific context of the status of a data link entity (Station or Bus Arbitrator).The T3 and T5 timers also refer to T0 and their lengths are calculated using T0 as a basis. T3 can be determined as $knT0$, where k is a coefficient that allows the BA's turnaround time to be covered and $n > 2$ is a function of the geographic address of the BA to which T3 is attached. T5 can be determined as twice of T0. T2 is not based on T0. This timer determines the length of a basic synchronized cycle. T2 is activated at the beginning of each basic synchronized cycle, when the BA takes on the status of emitting identifier frame. Since these timers have effect on the overall performance of the FIP network system, setting the values of timers is important.

2.2 Transmission Performance

FIP's transmission performance is very low because of its own protocol characteristics. FIP's efficiency can be calculated using transmission mechanism and frame coding.

The transmission mechanism of FIP is centralized one. An element known as the bus arbitrator (BA) controls the right of each data producer to the medium by emitting a frame containing an identifier. The FIP (WorldFIP, 1995) provides two types of data transmission services: variable exchanges and message transmissions and there are also two types of exchange service: periodic and aperiodic. The paper considers only the transmission performance in case of periodic variable exchange. Performance of other types can be also obtained in the same way.

Periodic transmission is automatically triggered by the communications system without the user requesting them. The BA broadcasts a variable identifier frame. A variable response frame is then broadcast by the sole producer of the information required. During this phase the information is taken from the network by consumers. When one transaction has been completed the BA begins the following transaction according to schedules defined when the system is configured. In other words, all periodic FIP transaction are made up of the exchange of two frames: an ID_DAT frame followed by an RP_DAT frame. The RP_DAT frame should appear within a given time. This time is called the turnaround time. Turnaround time is the time elapsed between the end of reception of one frame and the beginning of transmission of the following frame. Turnaround time (TR) is defined in the network management standard as $10T_{MAC} \leq TR \leq 70T_{MAC}$, where T_{MAC} corresponds to the time needed to transmit a physical layer symbol ('1', '0', 'V+', 'V-', 'EB+', or 'EB-'). Hence it is varied according to the transmission speed. This parameter can be very important in determining transmission performance. Fig. 1 shows the variable transmission sequence.

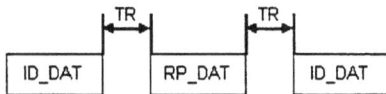

Fig. 1. Variable transmission sequence

Frame coding of FIP is carried out at each layer. Each layer adds service information to what it received from the layer above. The three layers add a total of 61bits, regardless of user data (WorldFIP, 1995). Fig. 2 shows an example of frame coding.

If n bytes data is transmitted the ratio of data to overhead of RP_DAT is $\frac{8n}{8n+61}$. Considering the usual size variable is about 2 bytes the data to overhead ratio is $\frac{16}{77} = 0.21$, which is very low. Furthermore the transmission mechanism requires more additional overhead. Since the transmission mechanism consists of one ID_DAT frame, one RP_DAT, and two TR, in which case we can

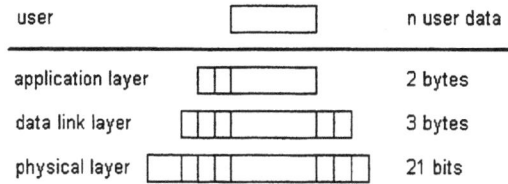

Fig. 2. Overhead of a frame

assume TR as $40T_{MAC}$ as the average value of $10T_{MAC}$ and $70T_{MAC}$. Then the efficiency in case that the transmission speed is 1 Mbps which means $1T_{MAC}$ is 1 μsec is $\frac{8n}{2*61+2*40+8n}$ and $\frac{16}{218} = 0.07$ in case of $n = 2$. This measure is too small to apply to industrial field but this is inherent to FIP. So we should tackle this problem if we want to use FIP more efficiently.

One solution is grouping variables of the same producer. Fig. 3 shows the procedure of scheduling of FIP. Grouping is important when the network should hold a number of variables. Assuming the network should transmit ten one byte variables every one msec then we should have the network whose capability can transmit $10 * (2 * 61 + 2 * 40 + 8) = 2100T_{MAC}$ every one msec. An approach is that we use a network the speed of which is more than 2.1 Mbps. However, if the number of variables increases there might be no solution. But if we can group ten variables in two groups as two five byte variables or one group as one ten byte variable, then the network has only to transmit $2 * (2 * 61 + 2 * 40 + 5 * 8) = 484T_{MAC}$ or $2 * 61 + 2 * 40 + 10 * 8 = 282T_{MAC}$, which is drastic decrement. So grouping of variables is strongly required in FIP application to improve the feasibility of scheduling.

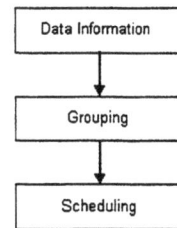

Fig. 3. Procedure of Scheduling

3. SCHEDULING OF DATA LINK LAYER

A distributed process control system is typically characterized by the presence of two kinds of processes; periodic and asynchronous ones. A periodic process such as sampling an analog signal from a sensor, performs highly repetitive operation at a fixed frequency. An asynchronous process, on the other hand, evolves in time in a way that cannot be foreseen a priori and it is affected by events occurring in the process itself (e.g., alarms) or outside it (e.g., a request from a remote process for the current value of a variable). In case of FIP the pre-run-time scheduling is carried out when communication medium is allocated to periodic variables. This means that the scheduling of periodic variables fixes the transmission time for each periodic variable in a scheduling table.

Medium allocation of FIP is fulfilled in data link layer and the allocation mechanism is basically centralized one. An element known as Bus Arbitrator controls the right of each data producer to the medium by emitting a frame containing an identifier. BA provides a synchronization function in order to guarantee the constant length of each cycle. Each type of transmission takes place in a specific window, i.e. respectively in a periodic window, an asynchronous variables window, an asynchronous messages window, or a synchronization window. These four windows constitute a basic cycle. So scheduling problem becomes making scheduling table of bus arbitrator which is known as Macro cycle which is composed of one or more than one basic cycles. Pre-run-time scheduling the paper considers is only periodic information because its period is predefined at the configuration stage while that of asynchronous information cannot be predicted. But there should be bandwidth for asynchronous information. Only periodic variable exchange is deterministic. This means that FIP can guarantee that a variable with a given period will be scanned at the proper instant.

3.1 Scheduling Constraints

The paper considers three kind of scheduling constraints. They are memory size, medium utilization, communication jitters, and uniformity of schedule. Memory size is one of important constraints in pre-run-time scheduling because if scheduling information to be stored is too big, it is not possible to store the information to the limited memory. In addition, in case of process control systems implemented by means of simple devices, they usually feature small memory size. So, when scheduling sequence is very long and it has to be stored in order to be repeated periodically, large amount of memory is needed. Moreover if we reduce the scheduling sequence, both the use of more complex methods than those usually used, and less efficient use of the transmission medium should be applied because of the possible need for redundant transmissions of a single produced value. Medium utilization is another because a too big utilization of periodic information in FIP is not desirable because the exchange of asynchronous information is not feasible if the utilization of periodic information is too big. The paper shows in section 2.2 that the transmission performance of FIP is very low which means the small data can cause utilization overflow. So utilization should be kept in mind in scheduling. The third constraints the paper considers here is

communication jitters. In some control application, jitters of control inputs and/or plant outputs due to communication network can have critical effect on the control system. So it can be an important constraint (Jae Jin Choi, 1994). Uniformity of schedule is the degree of distribution of scheduling result of periodic variables. A more uniform schedule of periodic variables can support a more uniform predictability of transmission of asynchronous variables. In the other hand a more biased schedule of periodic variables can support the transmission of a long asynchronous variables. In other words there is trade-off in uniformity.

In considering these constraints, the paper introduces the existing scheduling method and its problems briefly and suggests two kinds of scheduling methods. They are LCM-based method, reduction method, and jitters-resolving method. And the paper compare these method using a simple example. For each method two schedules are suggested. One is a more uniform schedule and the other is a more biased schedule. The example network consists of six periodic variables like Table 2. The bus arbitrator must scan six periodic variables. For each variable the arbitrator knows its period and its application type. Using the transmission time (here 1 Mb/s) and the turnaround time, the bus arbitrator can calculate the time needed for an elementary transaction made up of the transmission time for a question frame, followed by the transmission time for the associated response frame. The time column corresponds to the length of transactions, given a turnaround time of 40 μs.

Table 2. An example of FIP network

Variable	Period (ms)	Type	Time (μs)
A	5	INT_8	210
B	10	INT_16	218
C	15	OSTR_32	458
D	20	SFPOINT	234
E	20	UNS_32	234
F	30	VSTR_16	330

3.2 Existing Method (LCM-based Method)

LCM-based method is a common one in which the duration of basic cycle and macro cycle is determined as GCD and LCM of all periods of variables respectively. Since basic cycle is GCD of all periods scheduling is easily carried out according to the period of each periodic variable. However in cases of a small GCD or a too big macro cycle this scheduling method cannot be used. This method is very useful a small LCM and a big GCD. Fig. 4 shows a possible distribution of the identifiers on a time axis, as a function of the period of each of the variables. Each period of time constitutes a basic cycle. Basic cycle in this example are all of the same length: 5 ms. Macro cycle is composed of twelve basic cycles (60

ms), which requires a large memory space. It can be seen easily from Fig. 4 that the transmission time of some variables have jitters which can be ignored. The utilization of macro cycle in this case is $U_{LCM} = \frac{7724}{60000} = 0.129$ which is low utilization.

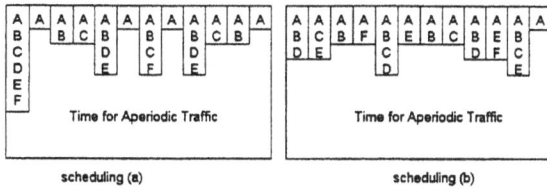

Fig. 4. An example of LCM based schedules

3.3 Suggestion 1 (Memory Reduction Method)

Reduction method is basically based on oversampling concept. This method give up utilization at the cost of memory space. First basic cycle is determined as GCD of periods of all periodic variables or some value equal to or less than the smallest period among all periods. Then the size of macro cycle is determined to meet the limited memory space. In this case macro cycle need not be LCM of all periods because the over-transmission of variables can compensate the periodicity. After determining basic cycle and macro cycle, medium allocation is made to make sure that more than one transmission of the variable is accomplished in the period of the variable. This method can be useful when memory constraints is strict. But this method can cause big jitters and utilization of communication media can be big. Fig. 5 shows a schedule of the example. Basic

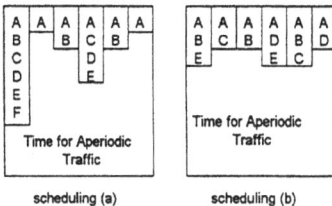

Fig. 5. An example of reduction method schedules

cycle in this example is determined as GCD of all periods (5 ms). Macro cycle is composed of six basic cycles (30 ms), which requires a reasonable memory space. It can be seen easily from Fig. 5 that the transmission time of some variables have communication jitters which cannot be ignored. The utilization of macro cycle in this case is $U_{Reduction} = \frac{4096}{30000} = 0.137$ which is rather low utilization.

3.4 Suggestion 2 (Jitters-Resolving Method)

Jitters-resolving method modifies the periods of periodic variables and makes them harmonic before scheduling, and then makes schedule using

LCM-based method. At first the basis period is determined which is usually the smallest period of periods of all periodic variables and becomes the duration of basic cycles. Then each period of each variable is determined as the largest value among the multiples of the basis period which are equal to or less than the period of the variable. This procedure improves the schedulability if only utilization permits. And then making schedule is carried out. This method provides pre-run-time schedule without jitters which means the distance of the consecutive transmission time of each periodic variable is equal. But this method can cause big utilization of communication media. Basic cycle and basis period in this example is 5 ms. Macro cycle is composed of four basic cycles (20 ms), which requires very small memory space. It can be seen easily from Fig. 6 that the transmission time of all variables have no communication jitters. The utilization of macro cycle in this case is $U_{J-resolving} = \frac{2990}{20000} = 0.150$ which is rather high utilization.

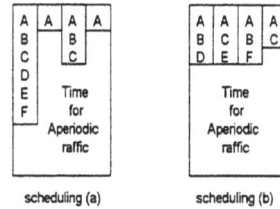

Fig. 6. An example of jitters-resolving method schedules

Using these results a good schedule which satisfy given constraints such as memory size, medium utilization, and communication jitters can be obtained.

4. GUARANTEEING THE VALIDITY OF THE INFORMATION

Validity of information such as time and space consistencies is another important problem (Jean-Dominique Decotignie, 1993). In industrial communication, information that has not arrived to its destination in the given time is often useless to the control application. Moreover the same piece of information may have different validity times depending on the user. It means that the network should provide a way for a user how to know the age of the piece of information. The problem lies in the difficulty how to decide whether an information is still valid or not. In some cases, the source of the information might send a new information that supersedes the first one and cancels the transmission of the first one. In some cases, the information that supersedes the first one might come from another source. So the communication network must be able to indicate if values transmitted are time and space consistent. Time consistency is mandatory for a good process and space consistency is required from a group of control devices. Space consistency is often ensured by special methods in distributed systems in which case one speaks about global state. So FIP introduced promptness and refreshment concepts to provide time and space consistencies information. Therefore the setting of promptness and refreshment timer is important.

The paper shows results for the two cases. The first one considers synchronization between DLL and APs which means that it assumes FIP can support some synchronization mechanism between of the two but this assumption is not specified in the current FIP specification. The other one doesn't put any assumption to the current FIP specification. The results below use the symbols to represent timing relations in the AP to be schedule. They are Phase (ϕ_i) which means the starting time of a synchronization variable i, deadline (d_i) which means the time the AP should be completed to, period of application using the synchronization variable i ($T_{ap,i}$), and T_e which represents the basic cycle of DLL.

In the first case, the paper suggests that $\phi_i = t_{max}$, $d_i = T_i - t_{max}$, $T_{ap,i} = T_i$, refreshment timer value $T_r = d_i$ (producer), and promptness timer value $T_p = d_i$ (consumer), where i is a variable, T_i is period of variable i, T_e is elementary cycle, t_{max} is maximum periodic data transmission time per elementary cycle T_e. The scheduling results of communication interface is shown in Fig. 7.

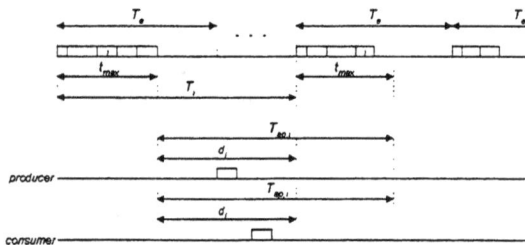

Fig. 7. The schedule of application processes

In the other case, the paper suggests that the period of producing AP $T_{ap,i}$should be a half of the refreshment timer value T_r and the peirod of consuming AP should be a half of the promptness timer value T_p. refreshment timer T_r should be twice the period of AP, and promptness timer T_p should also be twice the period of AP. From this result it is apparent that if the twice oversampling of AP is performed there is time and space consistencies.

5. CONCLUSION

The paper dealt with the real time properties of FIP such as timers, transmission performance, scheduling methods of DLL, time and space consistencies of application processes. The paper recommended the setting value of timers of data link layer and showed the transmission performance. The paper suggested that since transmission performance is low inherently the grouping procedure of information is required before scheduling procedure.

The paper also suggested how to schedule the communication media to meet the system constraints. They are memory reduction method and jitters-resolving method. In doing so a guideline for choosing one of them according to the objective and constraints of the system is suggested naturally. The constraints considered in choosing one scheduling method are memory size, communication medium utilization and jitters. Since the constraints are substantial the guideline helps when fieldbuses are applied to real industrial fields.

In addition, the paper also proposed a real time mechanism at application process level to guarantee real time which is typical obstacle when fieldbuses are applied to industrial fields. The paper suggested two real time mechanisms to guarantee time and space consistencies of application processes. One considers the case when there is synchronization mechanism between data link layer and application layer, the other when there is no synchronization mechanism between both. The paper also shows the performance of the two cases.

6. REFERENCES

Ching-Chih Han, Kwei-Jay Lin (1992). Scheduling distance-constrained real-time tasks. *Proc. of Real-Time Systems Symposium, IEEE* pp. 300–308.

Jae Jin Choi, Seung Ho Hong (1994). Determination of control device sampling times in the fieldbus systems. *Proceedings of ASCC'94*.

Jean-Dominique Decotignie, Patrick Pleinevaux (1993). A survey on industrial communication networks. *ANN. TELECOMMUN.*

Jia Xu, Dvaid Lorge Parnas (1991). On satisfying timing constraints in hard-real-time systems. *ACM SIGSOFT '91* pp. 132–146.

P. Lorenz, Z. Mammeri (1995). Real-time software architecture: Application to fip fieldbus. *AARTC '95, IEEE* pp. 415–423.

P. Pleinevaux, J.D. Decotignie (1988). Time critical communication networks: Field buses. *IEEE Netowrk.*

S. Cavalieri, A. Di Stefano, O. Mirabella (1995). Pre-run-time scheduling to reduce schedule length in the fieldbus environment. *IEEE Tr. on Software Engineering* **21**(11), 865–880.

WorldFIP (1995). *General Purpose Field Communication System, prEN 50170*. WorldFIP.

INVESTIGATION AND IMPLEMENTATION OF FIELD COMMUNICATION OPEN POLICY

Minrui Fei , Wen Liu, Yunchao Qiu, Wenpeng Lang, David Guosen He

School of Automation, Shanghai University
Shanghai 200072, P.R.China

Abstract: This paper presents latest research, development and realization of the Field Communication Open Policy (FCOP) during latest fast and strong development of open protocol tendency of lowest layer network (fieldbus) for factory automation, CIMS, DCCS and PLC systems etc. nowadays. The FCOP and its support techniques consist of two main parts: a) the technique of Graphic-oriented Programmable Protocol Interface(GPPI); b) the technique of multiprotocol configuration Field Communication Control Device(FCCD). The practical operation shows that FCOP has correct philosophy and realistic policy.

Keywords: Distributed computer control systems(DCCS); fieldbus; communication protocol; configuration; programming environments

1. INTRODUCTION

Field Communication(FC) which includes the fieldbus is prevailing widely for factory automation(FA), building automation(BA), laboratory automation(LA) etc. During latest 20 years more international authorities such as ISO, IEC, IEEE, CCITT. etc. had delivered a series of FC standard protocols which may be suitable for FCs, and many famous computer companies and automation instrumentation companies have almost their own patented protocols respectively thus these situations lead to great troubles for the conceptions of open field communication protocol.

FCOP is one policy which can support the open field communication protocol, and has the functions of building patent field communication protocols and configuring different types of field communication protocol, to be interconnected with different standards and patents' field communication protocol systems.

FCOP can meet not only those international field communication protocols which are either opened or unopened as the present status, but also it can reduce of or refrain from complex and repeat labors of R&D of patent communication system in order to be advantageous to FA, BA, LA etc. single or multiprotocol communication systems' construction and to field distributed computer networks connections of CIMS, CIPS, DCCS, PLC etc.

Realistic FCOP function R&D consists of following points:

1) basic hardware structures and software modules of novel multiprotocol Field Communication Control Device(FCCD) for integrated different types of communication equipment and

supporting different communication protocols;

2) methods of realizations for present international open or standard field communication protocols;

3) Graphic-oriented Programmable Protocol Interface(GPPI) for patent communication protocol of field intelligent equipment, and its general hardware support techniques;

4) software and hardware configuration techniques for realization of different communication structure and system in field intelligent equipments;

5) user interface of general microcomputer and multiprotocol configuration FCCD;

6) practical simulation techniques to complete the system performance estimation through simulation models of single or multiprotocol field communication system.

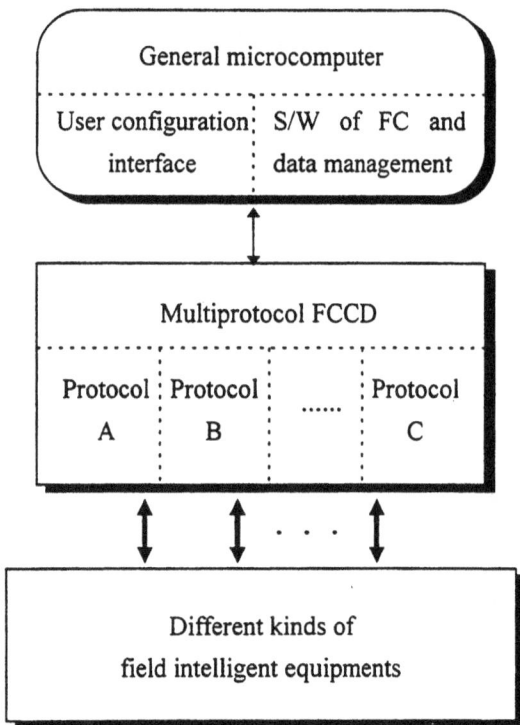

Fig. 1. Schematic diagram of the system constructed by FCOP

2. GRAPHIC-ORIENTED PROGRAMMABLE PROTOCOL TECHNIQUE

Because of incomplete open of field communication so there are many kinds of patented protocol in different countries, locals or groups and many companies nowadays. Even the open field communication protocols have been finished however it requires some time of practical implementation and examination finally by different manufactures and users. Therefore, based on multiprotocol configuration FCCD, R&D of GPPI supports more kinds of patent protocol generation.

Definition 1: GPPI means the graphic symbols used for description of field communication protocol's basic loop and process, and the availability for operators with generated interfaces of nonprofessional computer type programmable protocol's flow process.

GPPI utilizes some concepts of the Petri Net and treats their practicality. Owing to greatly reduce requirements of user's computer software programming and decreasing the strength of works for field communication protocol model, it will benefit to the practicality and interoperability realizations of miltiprotocol field communication system and its extension for development and construction DCCS.

2.1 Idea of design

FC belongs to the lowest layer network of computer communication system. Generally FC only uses three layers of ISO/OSI reference model which has data link layer as the key of field communication protocol and the core of GPPI.

In order to realize the general and formal description of each loop, then to extract the basic elements of field communication protocol from standard protocols thus the standard and patent protocols are generated by these configurations of elements. Owing to Petri Net's special features of graphic symbol, layer construction and ease of integrated system (Kamath, et al., 1986), the idea of design partly uses reference of Petri Net's concept and method for extracting the basic communication elements. However the final goal is different because the Petri Net lies generally in system performance evaluation and operation mechanism analysis (Zhou, et al., 1991), and GPPI is used to describe generated field communication protocol software flow process, eight basic communication elements and their formal graphic symbols(shown in Fig.2) are redefined as following:

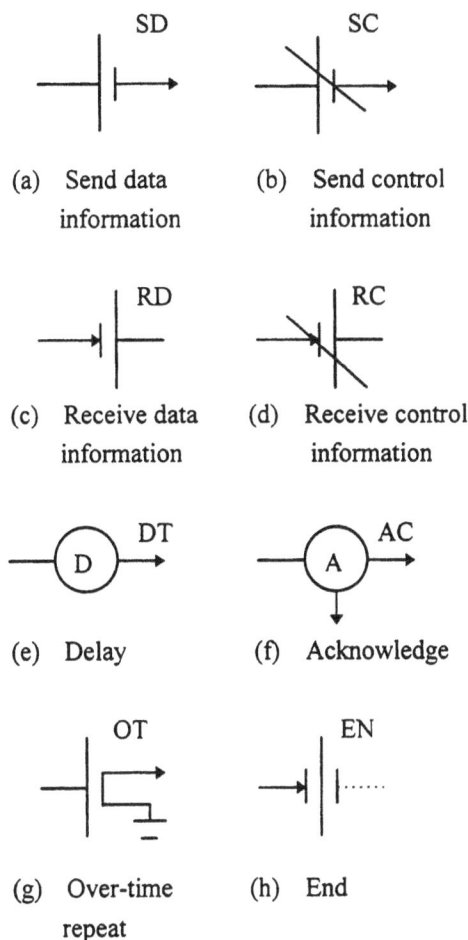

Fig. 2. Basic graphic symbols in GPPI

1) Send data information(SD)—to send user's data package;

2) Send control information(SC)—to send calling, response signal, address, time set and synchronous control signals etc. from protocol;

3) Receive data information(RD)—to receive user's data package;

4) Receive control information(RC)—to receive calling, response signal, address, time set and synchronous control signals etc. from protocol;

5) Delay(DT)—the waiting measures adopted by two sides of communication for the synchronous speed of transmission;

6) Acknowledge(AC)—to decide the protocol flow process according to addressing control, access control and error control;

7) Over-time repeat(OT)—to repeat deliver until success or failure for several times when over-time and no response with abnormal exit;

8) End(EN)—the normal exit in present communication status.

As to each basic communication element, it needs definition of its basic data structure as following except above mentioned graphic symbols:

☞ Function code: the communication function represented by certain concrete numerical notation;

☞ Flow process code: the order and position of basic communication element in communication protocol;

☞ Correct or wrong flow direction: the branch and direction of the flow process according to communication response;

☞ Auxiliary parameter: the control information and parameter included in the basic communication element.

The following illustrative example of simple patent protocol used for field intelligent equipment is shown by means of GPPI configuration method and principle. The example has following basic content and orders of A-user field communication protocol:

I. to transmit the address information and to get synchronous response;

II. to decide the response: if it is correct then to next step, i.e. data information transmission; if it is wrong then repeat step I. again until the correct or response times over regulated limits;

III. to transmit data information and to get error control response;

IV. to decide whether the step III. error control correct or wrong. If it is correct then communication exit; if it is wrong then repeat step III. until the communication will be correct or repeat times over regulated limits then declare its failure.

According to non-formal description of above mentioned example then to use formal graphic symbols description of basic communication element as following Fig.3.

Each basic communication element in Fig.3 needs a flow process function table which should define each parameter respectively in order to establish complete data structure for final correct generation of field communication protocol and its actual code in graphic-oriented programmable protocol interface.

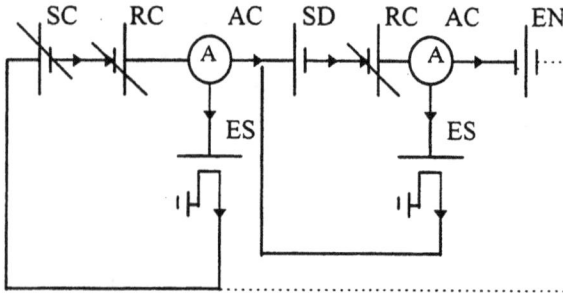

Fig. 3. Illustrative example of formal graphic symbols description of field communication protocol

2.2 Realization of software

The graphic-oriented programmable protocol technique used for field communication protocols is realized by its software interface and following three submodules(shown in Fig.4) will mainly support its software of structured function:

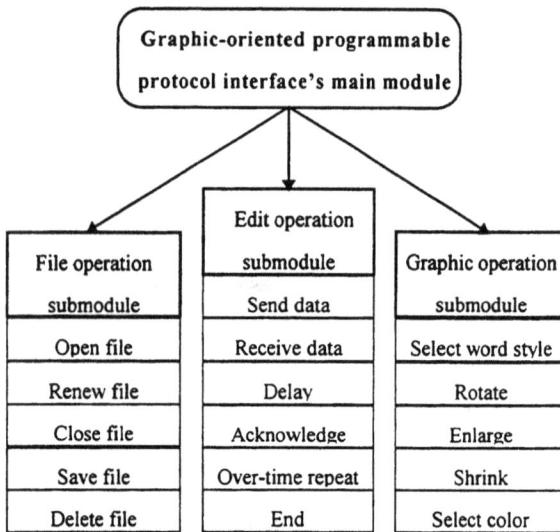

Fig. 4. Module schematic diagram of graphic-oriented programmable protocol interface

1) Edit operation submodule of field communication protocol—This is the core of complete graphic-oriented programmable protocol interface software. It supplies operator with a series order items of basic communication element. The operator can select certain item through the menu and the cursor will correspondingly change the graphic symbol required then hints selected item functions.

Hence the operator can move the cursor within full position of complete screen by fingering its definite position, to knock the mouse once more toward leftkey and the flow process function table will be pushed thus to complete corresponding data structure function of basic communication element.

2) File operation submodule—This is used for the operator taking out his edited field communication protocol file. It is partially similar to file operation of WINDOWS: it can build any new field communication protocol file, take any editing or required edit of field communication protocol file, and it can open simultaneously more field communication protocol files. Especially for store operation of field communication protocol files, it can treat edited field communication protocol files and rearrange in series orders for communication flows. Therefore it can cancel all symbols and data structures which are not related to field communication protocol thus the capacity of files is reduced, source of microcomputer is saved and the goal of increasing process speed is attained.

3) Graphic operation submodule—This submodule can satisfy description of formal graphic symbols and edit field communication protocol for supply auxiliary function and tools such as enlarge, shrink, rotate, select word-style, color selection etc. in order to make complete graphic-oriented programmable protocol software beautiful, vivid pictures then fulfills the visual requirements of the operator.

3. MULTIPROTOCOL CONFIGURATION FIELD COMMUNICATION CONTROL TECHNIQUE

The multiprotocol FCCD will realize the important material base for FCOP. Its main design is used for process control and manufacture automation etc. communication management of industrial fields nowadays. This FCCD plays the middle role between field equipments and microcomputer of control system and connects more field intelligent equipments such as instrument and device with CPU thus build up the lowest layer networks of industrial control field. The multiprotocol FCCD consists of self-difined

94

system bus based multi-CPU parallel processing system which can support single or multiprotocol field communication system. This FCCD can have standard and open protocols which are composed by communication chips but patented protocol function is finished by configurations by means of byte, bit and

character-oriented communication protocol which are supported flexibly by general communication chips.

3.1 Support of hardware

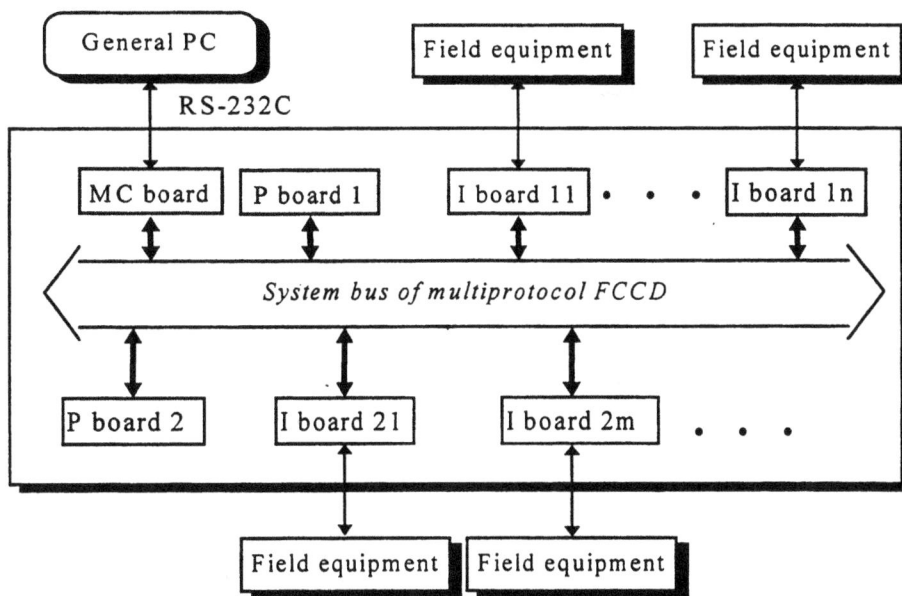

Fig. 5. Hardware construction black diagram of miltiprotocol FCCD

The multiprotocol FCCD's hardware consists of mother-board and insert-board (Fig.5 shows their construction block diagram). The former supplies different kinds of D.C. sources and extended slots and fix-frame etc., and self-defined system bus is also built with mother-board and extended slots' signal wires. The latter consists of three kinds, i.e. main control(MC) board, protocol(P) board and interface(I) board which shows first two boards have attached CPU used as general PC inserts as feasible as possible upon any extended slot. The system bus will do all signal connections between different insert boards, control and power supply.

The main control board can manage system bus of miltiprotocol FCCD and realize interoperation and communication to general microcomputers by means of interface of RS232C or connection of MODEM. Besides, main control board can make information exchange through bi-port RAM of each protocol board.

The protocol board is the key board of miltiprotocol

FCCD because all information communication of field intelligent equipments depends on it. The protocol of communication and the size of RAM can be selected and configurated by the operator according to concrete field situations. Generally one common protocol board can configurate more protocols but present developed protocols of data link layer have the serial asynchronous communication protocols of BSC, HDLC/SDLC, DDCMP etc. Although one protocol board can only fulfill one communication task, there are many protocol boards up to ten thus they can solve the trouble of multiprotocol parallel jobs. There are several specifications selected from 8K, 16K, 32K, 40K, 64K etc. of RAM space through wire jumper.

The interface board is controlled directly by the protocol board therefore it realizes physical layer interface between miltiprotocol FCCD and field intelligent equipments. The quantity and capacity of interfaces on the interface board can be selected by the operator according to realistic situations. Now there are many developed interface boards such as

RS232C, RS422A, RS485A. One interface board can supply eight RS232C interfaces or six RS232C and two RS485A interfaces; or seven RS232C and one RS422A interfaces. Owing to one protocol board can manage eight interface boards thus one protocol board can connect more than 300 physical nodes thus whole multiprotocol FCCD can further extend its physical layer capacity. But from the point of view of distribution risk design, nowadays there is a regulation to limit maximum 256 logic nodes for one set of multiprotocol FCCD.

3.2 Software system

The software related to multiprotocol FCCD can be divided into three parts: user configuration software, main control board software and protocol board software. The first part will be realized by high language of general microcomputer, and other parts can be realized by general or special assembler of multiprotocol FCCD. Each software adopts modular programming method in order to extend functions of software system and to be easy of system maintenance.

The generated system information, the selected node type information and the set parameter information in general microcomputer user configuration interface will be transmitted to main control board of multiprotocol FCCD through serial interfaces. Simultaneously, all nodes' data packages and system status information of multiprotocol FCCD will be concentrated on the main control board then to general microcomputer. Within the multiprotocol FCCD the main control board and protocol board can make information exchange through system bus and bi-port RAM on the protocol board. The former can send related nodes' parameter settings into protocol board; on the other hand, the protocol board's system status information and nodes' data package will be delivered to main control board.

The configuration functions of multiprotocol FCCD are shown concretely as following: the operator can set conveniently whole field communication system by microcomputer which includes node numbers, node communication address, node communication protocol, communication interface, communication

bode rate and node communication data types etc. according to concrete situations of field intelligent equipments.

By combination of software and hardware configuration, the operator can build an open or different patents multiprotocol field communication system conveniently and flexibly. Thus this FCCD user and every integrated company need not to research special FCCD which leads to save great man, material and money from macro point view.

4. CONCLUSIONS

The FCOP can meet the case of automation development, tendency of international system development and primary status of P.R.China nowadays. Low cost DCCS (Fei, et al., 1996) with our laboratory self-design and self-made Model FC-11 instrument, Chinese Model 861 and CP-20 instrument etc. for field control unit has been built, and our intelligent control and real-time data management of electrical furnace are finished successfully.

On conclusion, the practical operation shows that FCOP has correct basic philosophy and realistic policy but further considerations and more complete efforts will be need in field practices in near future with our optimistic prospects.

REFERENCES

Fei, Minrui and et al.(1996). Low Cost DCCS and its Functional Design. *Elec. Automation Joul.,* **18**, 40-42.

Kamath, M. and N. Viswanadham (1986). Applications of Petri Net Based Models in the Modelling and Analysis of Flexible Manufacturing Systems. *Proc. of IEEE International Conference on Robotics and Automation*, pp. 312-317

Zhou, Mengchu and F. DiCesare (1991). Parallel and Sequential Mutual Exclusions for Petri Net Modelling of Manufacturing Systems with Shared Resources. *IEEE Trans. on Robotics and Automation*, 7, 515-527

AN EXTENDED TCP/IP PROTOCOL OVER THE LOCAL AREA NETWORK FOR DCCS

Jaehyun Park, Youngchan Yoon, and Sangchul Lee

Department of Industrial Automation,
Inha University,
Incheon 402-751, Korea
email:jhyun@rcsl.inha.ac.kr

Abstract: This paper proposes an extended TCP/IP protocol over local area networks, that can be used for soft real-time systems including distributed computer control systems and manufacturing automation systems. Because the proposed protocol extends the standard TCP/IP, all the existing application softwares can be used with the proposed protocol without modification. The proposed TCP/IP also provides the periodic transmission mode(PTM), which provides a very efficient transmission method of the periodical data for real-time monitoring and control systems. With PTM, the periodical data collection and updating the control signals are possible with relatively small traffic overhead. This paper includes the computer simulation results, and the prototype system has been applied and evaluated to the large scale distributed control systems and human-like robot systems.

Keywords: Ethernet, real-time communication

1. INTRODUCTION

Ethernet and TCP/IP pair is one of the most popular protocols for the MAC(Medium Access Control)(Metcalf and Boggs, 1976) and the middle range protocol layers in industry. Because TCP/IP is ported and used on hundreds of operating systems and hardwares, there are lots of system-independent applications developed on the top of TCP/IP. Although TCP/IP on the top of Ethernet works well for the ordinary network applications such as remote login, file transfer, and remote procedure call, this is hardly used for the real-time applications such as manufacturing automation and distributed computer control applications. The major difference between the general internet applications and real-time applications is the time constraints in the message delivery. The usual internet applications permits relatively large amount of variances in the message delivery time and throughput. The real-time applications, however, have strict or very tight requirements in

the message delivery time. Failure to meet these requirements may cause a severe trouble in the overall system. Although a sophisticated real-time network protocols can be used to meet these strict real-time constraints, general softwares can not be used on the top of these new real-time network protocols without modification. This paper focuses an approach to use the popular TCP/IP - Ethernet protocols for the soft real-time systems with keeping compatibility.

There are many researches to use Ethernet for the real-time applications. Because the packet collision of Ethernet is one of the major reasons why it can hardly used for real-time application, most of the researches concentrate on solving the packet collision problem (Tanenbaum, 1989). BRAM(Broadcast Recognizing Access Method) (Chlamtac et al., 1979) and MBRAM(Modified BRAM) (Signorile, 1988) are the protocols that resolve the collision problem in Ethernet network. Although, these protocols are collision-free protocols and provide the deterministic latency, there are several reasons to prevent these collision-free MAC protocols from being widely used.

- No standard contention-free MAC protocol is fixed yet. The manufacturer of control devices and softwares are not willing to support non-standard protocols. They may want to support general solution based on world wide standard such as CSMA/CD, token ring, token bus, TCP/IP, and etc. Backward compatibility must be supported so that the existing software should be reusable.
- Reprogramming the I/O drivers is required for all the existing Ethernet devices. In case of the embedded MAC drivers, such as MC68EN360 CPU, it is impossible to modify MAC layer without redesign the chip.
- All nodes connected to physical media should support the same scheduling algorithm. Even single node that use standard MAC protocol may cause unpredictable congestion problems. Even though a special conflict-free protocol such as PCSMA (Yavatkar, 1994) is developed, most of the MAC protocols can not coexist with each other.

These reasons make it difficult to use modified MAC protocol for the real-time applications.

Another approach to use TCP/IP - Ethernet protocols for the real-time applications is to improve the performance of TCP/IP layer. TCP/IP was originally designed to be used in the internetworking environment where long latency and relatively unreliable data transmission are expected. To cope with this unreliable connection due to inter-networking, multiple steps of error controls are provided. To maintain optimal utilization of the low bandwidth channel, standard TCP/IP uses the best-effort paradigm for packet scheduling, buffer management, feedback, and end adjustment(Lefelhocz et al., 1996). Recently, a series of algorithms and implementation techniques improving the performance of the transmission control protocol(TCP) have been proposed: TCP for the transactions(ISI, 1994; ISI, 1992), receiving overhead reduction technique(Clark et al., 1989), and header prediction algorithm(Jacobson, 1990). While these researches are useful for the conventional internet applications, they are not efficient for the real-time applications over Ethernet-only environment that most of the real-time systems are used in.

This paper proposes a modified transmission control protocol(TCP) to reduce communication latency over Ethernet link without modification of MAC layer. For this, this paper introduces another type of service called least-effort. Even though least-effort is not able to be applicable to the low bandwidth link, internet, it shows a relatively high performance for a special purpose network such as manufacturing networks based on Ethernet that has a higher bandwidth than ordinary internet. This paper also proposes a special service called PTM(Periodic Transmission Mode). This mode increases utilization and results in lower transmission delay.

Next section describes the key features of the proposed extended TCP protocol. Implementation issues and evaluation results follows in the subsequent sections.

2. KEY FEATURES OF LAN-TCP

This paper extends the standard TCP protocol for the Ethernet-only environment, and the proposed protocol is called LAN-TCP(TCP for local area network only). The LAN-TCP has three major features : least-effort strategy, structure oriented protocol, and periodic transmission mode. In this section, these features are described in detail.

2.1 *Least-effort strategy*

The standard TCP follows the best-effort strategy to transmit a packet over a physical network link that is usually slow. TCP sliding window is a good example of explaining best-effort delivery (Comer, 1995; Stevens, 1994). It makes stream transmission efficient in wide area network, and it is an important feature of TCP for internetwork between high speed network and lower one. The best-effort strategy, however, increases the possibility of packet collision when used with high speed link like Ethernet. This packet collision is one of the major reasons of unexpected latency of packet delivery over Ethernet. To reduce this packet collision without modification of MAC layer, the least-effort strategy, contrary to the best-effort strategy, is proposed in this paper. The main idea of the least-effort protocol is that a certain limitation of the network bandwidth is reserved to single node, and the operating system is responsible for managing this limitation. The bandwidth limitation reserved to each node depends on the system architecture, the operating systems, and the transfer rate of media. The minimum bandwidth can be set to default or required amount requested by applications when applications open or bind SAP(Service Access Point) of transport layer or socket layer in BSD socket compatible environment. To maximize the effect of the least-effort strategy, LAN-TCP does not use the sliding window algorithm that is derived from the best-effort paradigm and developed for the inter-networking purpose. In the LAN-TCP, to disable the sliding window, the window size is limited to MSS(Maximum Segment

Size:(Comer, 1995; Stevens, 1994)). If one of two machine has a LAN-TCP facility, bandwidth limitation can be achieved by pseudo window size which is described in the next section.

Least-effort strategy is applied only to TCP or socket layer in LAN-TCP. There are four reasons for this restriction.

- Port de-multiplexing is applied only above the IP layer and the output data of IP layer are transmitted or queued to IP buffer without further de-multiplexing. Generally IP fragmentation is not required in local area network because TCP process already knows the size of MTU(Maximum Transfer Unit). If the bandwidth is not managed by TCP process, de-multiplexed TCP segments stored at IP buffer must be re-arranged to distribute transmission rate of each application processes. This rearrangement will be another overhead of protocol layering and make the buffer management complex.
- Application process should be pended until sending system call is terminated if multiple segments are sent, and TCP process passes segments to the IP process at intervals for pending duration.
- LAN-TCP should be compatible and coexistable with TCP. If least-effort delivery is not required, the standard TCP connection should be explicitly used. IP can not negotiate connection type with remote machine due to its connection-less feature. Compatibility is an important feature of LAN-TCP. If LAN-TCP can not support conventional TCP protocol, the problem of difficulty of implementation remains same as collision-free MAC protocol.
- IP or TCP process must know which application process is sending packet and how much bandwidth is reserved. IP process can be modified to deal with these information, but TCP process already has these facility known as TCB(Transmission Control Block) (Comer and Stevens, 1994; Wright and Stevens, 1995). TCB can be easily modified to implement least-effort mechanism.

2.2 *Structure oriented protocol*

TCP is a stream oriented protocol, and TCP's stream is unstructured. This unstructured stream makes programming inefficient. On the contrary, UDP uses structured datagram delivery, but UDP doesn't have transmission control facility. Application process using UDP is responsible for controlling transmission. LAN-TCP is datagram and reliable stream oriented protocol. In the LAN-TCP, stream is not divided into bytes but struc-

tured datagram. The destination machine passes to the receiver exactly the same sequence and size of blocks that the sender passes to it on the source machine. This feature makes buffer management of LAN-TCP easy and structural.

2.3 *Periodic transmission mode*

The periodic data transmission is one of the fundamental communication method for the real-time systems including distributed computer control systems. When a packet is transmitted periodically, the receiver is able to predict next arrival time in local area network. LAN-TCP supports a special transmission mode for the periodic transmission called PTM(periodic transmission mode). To identify the type of segment data, TCP header contains 6-bit fields: URG, ACK, PSH, RST, SYN, and FIN. Each bit is used to initiate transition of TCP state machine(Comer, 1995; Stevens, 1994). Among them, ACK bit is important for the PTM mode. To the TCP acknowledgment scheme, ACK is sent by destination machine after segment from source machine was received. Although destination machine can generate only last ACK for acknowledgment of multiple segments in sliding window algorithm, the receiver always generate ACK when segment is transmitted periodically because each segments are transmitted every the fixed interval. Generation of series of ACK can be avoided in PTM mode. ACK packets are not sent after PTM connection is established. ACK packet is used only for negative acknowledgment when destination machine didn't receive segment until next arrival time. From the queuing analysis of CSMA/CD, PTM can reduce the media traffic up to half or less(Rom and Sidi, 1990; Tobagi and Hunt, 1980).

3. IMPLEMENTATION

This section describes major issues to implement the LAN-TCP. The various techniques for optimizing TCP is not implemented in the prototype LAN-TCP yet.

3.1 *Connection management*

Only a slight modification of standard TCP is required for implementing the LAN-TCP. The connection establishing and closing scheme of LAN-TCP is almost same as that of the standard TCP except the PTM mode. Generally, TCP options follow only SYN segment while the connection is established. Standard TCP header contains 6 reserved bits for future use. Two unused bits are used for negotiating connection of LAN-TCP.

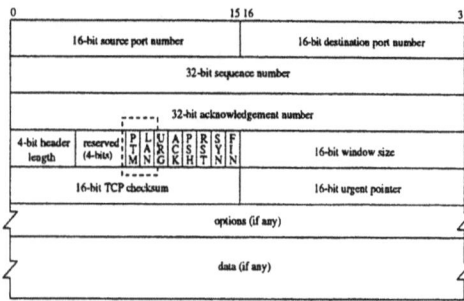

Fig. 1. LAN-TCP header

One is for LAN-TCP connection and the other is for PTM connection. The position of two bits is shown in Fig. 1, enclosed by dashed box. 'LAN' is used for LAN-TCP connection and 'PTM' is used for PTM connection. According to IAB internet standard, these bits must be zero(ISI, 1981).

There are some congestion control mechanism for best-effort delivery such as slow start algorithm, round-trip time measurement, Nagle algorithm. These do not need to be implemented in LAN-TCP because LAN-TCP is only for local area network especially inside subnet.

3.2 Additional timers for LAN-TCP

TCP process requires various timers to control transmission: retransmission timer, persist timer, and keep alive timer(Wright and Stevens, 1995). Some of them are disabled while LAN-TCP connection is established. Besides these standard timers, two additional timers are added for LAN-TCP: in-band timer and PTM timer. In-band timer is used by sender side for bandwidth limitation, and PTM timer is used by receiver side to send ACK for negative acknowledgment while PTM connection is established. Before a TCP segment is sent, TCP process must check whether in-band timer is expired or not. If in-band timer is already expired, segment is sent immediately, otherwise, TCP process is delayed until the timer is expired. After a segment was sent, in-band timer is initiated. The operation of two timers are depicted in Fig. 2 and Fig. 3. If PTM connection was established, all of TCP timer must be disabled in PTM connection requester except keep alive timer, in-band timer, and PTM timer. In most operating systems, the resolution of TCP timer is around several tens of mili-second unit. This resolution is sufficient to manage control mechanism of the standard TCP, but higher resolution is required for LAN-TCP. It is somewhat easy to obtain higher resolution in RTOS(Real Time Operating System). POSIX standard supports a function 'nanosleep' which has higher resolution, but real resolution is bounded up system clock and depends on operating system and processor. To fix this variation, the resolution of LAN-TCP timer is

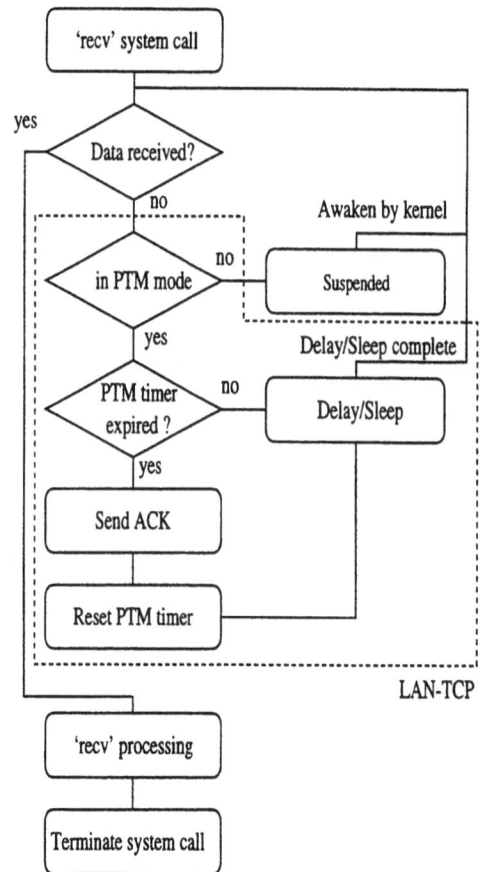

Fig. 2. LAN-TCP recv processing

set to the system clock. With this resolution, the maximum bandwidth depends on system clock. For example, if system clock is 60Hz and media MTU is 1500, then burst transfer rate of single process can be up to 90 kbytes.

3.3 Periodic transmission mode

The PTM connection is not bi-directional. When LAN-TCP connection is established, only sender set 'PTM' bit in SYN packet, thus uni-directional PTM connection is established. Transmission interval is automatically detected by receiver side, and it decreases from initial value to mean value gradually.

In PTM mode, PTM timer notifies that next packet is not arrived yet, and ACK, which means negative acknowledgment in PTM mode, must be sent(see Fig. 2). If sender side receives ACK, sequence number of ACK is checked, and corresponding segment must be sent again. Standard TCP process already has a facility, retransmission buffer, enough to solve this problem. Mostly ring-style buffer is used with retransmission buffer. In typical TCP process, acknowledged segment in retransmission buffer is removed, but it must be left in PTM mode. Segment in ring buffer is not removed but destroyed by head pointer of ring buffer step by step. If receiver side can not process

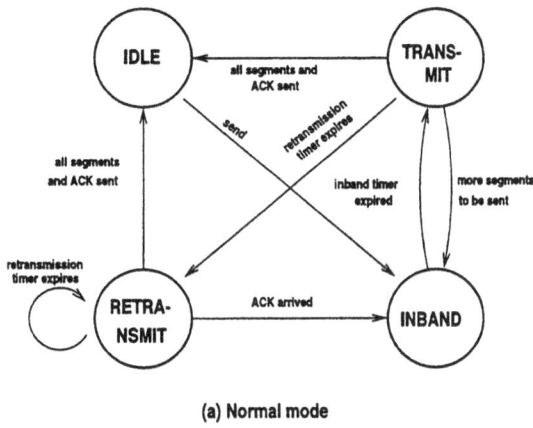

(a) Normal mode

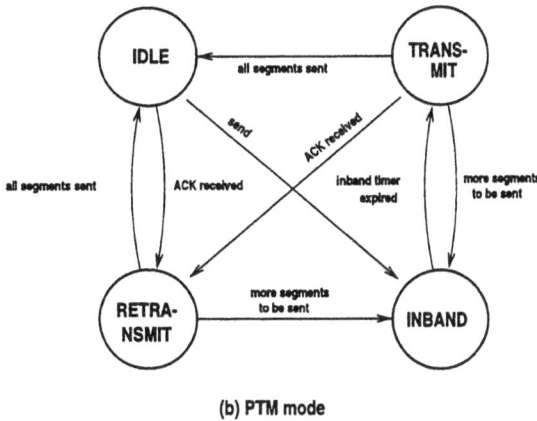

(b) PTM mode

Fig. 3. finite state machine used for LAN-TCP send processing

periodic packets in time by unexpected disturbance, then series of segments can be dropped. If dropped packets are out of ring buffer, these packets can not be retransmitted. In this case, the receiver must send source quench ICMP message, or the connection will be aborted. ICMP is only for internet protocol, but it is the only way to notify alert status.

3.4 Backward compatibility

Because the LAN-TCP is compatible with the standard TCP, the connection between the LAN-TCP nodes and the ordinary TCP nodes can be established, and two protocols work well without conflict at all. TCP has a facility known as window size advertisement to manage sliding windows. Window size in TCP header(see Fig. 1) is the size of free TCP buffer in remote machine's TCP process. Local machine sends full size of window which can be received and stored in remote machine. To prevent remote machine from sending more than one segment of window, pseudo window size is used in LAN-TCP. If one of two machine adopts LAN-TCP, bandwidth limitation can be achieved easily by pseudo window size. LAN-TCP uses pseudo window size rather than actual window size when the remote machine doesn't support bandwidth limitation. The maximum pseudo

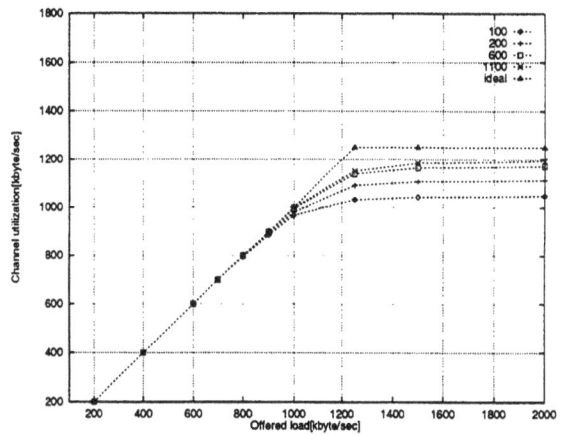

Fig. 4. Evaluated Performance of CSMA/CD with 20 nodes

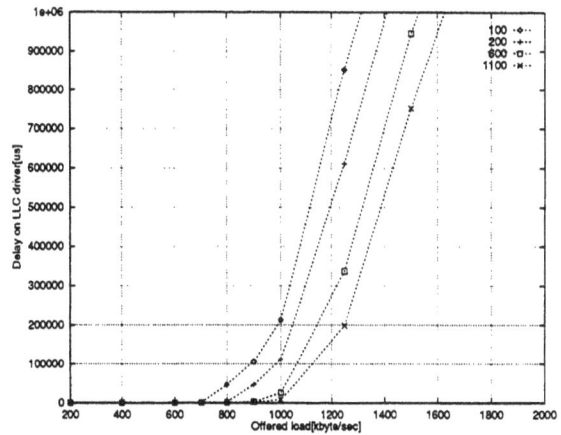

Fig. 5. Delay on LLC with 20 nodes

window size will be limited to MSS(Maximum Segment Size).

If applications use PTM, they need to use special socket system call rather than standard socket system call. These applications pass parameters different from normal LAN-TCP connection when create the end point of connection. For example, 'socket' system call can be called with new parameter 'SOCK_PSTREAM' rather than 'SOCK_STREAM', if PTM connection is required. If remote machine doesn't support LAN-TCP, PTM connection can not be established, thus the connection will be aborted. In this case, normal LAN-TCP connection is established and works well without conflict with conventional TCP.

4. PERFORMANCE EVALUATION

To evaluate the performance of the proposed protocol, a computer simulation package is developed. In the CSMA/CD protocol, if two stations detect an idle state and begin transmission at the same time, a collision will occur. Any station detecting a collision stops its transmission and generates jam signal, and then waits a random

101

period of time before retransmission. To avoid overloading the channel, the range of retransmission interval increases using a binary exponential back off algorithm. The complete analysis of binary exponential back off algorithm is very complex. Even though the simplified models were proposed(Tobagi and Hunt, 1980; Boggs *et al.*, 1988), the precise model with complete binary exponential back off algorithm should be used to evaluate the performance of the proposed protocol. Additionally, to evaluate the effect of physical environment, simulation model should contain a facility of handling propagation delay.

The first simulation is evaluating the basic utilization under heavy load to validate the computer simulation package itself. Utilization of the CSMA/CD shown in Fig. 4 is almost same as the measured utilization by Shoch and Hupp(Shoch and Hupp, 1980). Evaluated utilization is up to 95% with 20 nodes and 1100bytes of packet length. Average transmission delay from network layer to the destination node is shown in Fig. 5.

4.1 Simulation model

The network configuration simulated consists of four components(two periodic servers, one burst server, and one burst client), and two groups of periodic sources. Periodic server receives packets from the periodic data sources and sends acknowledgment packet to them in standard TCP model, but in the LAN-TCP model, acknowledgment packet is not sent.

Burst server and burst client exchange series of segments and acknowledgment packets as sliding window scheme. Burst client in this simulator does not generate packets by probability function but follows actual transmission scheme which exchanges one acknowledgment packet per several segments. Burst client does its best to transmit packets as soon as possible in conventional TCP model, but in the LAN-TCP model, transmission interval is never exceeded a preset limitation reserved at each node.

4.2 Simulation results

Fig. 6 through Fig. 8 show a delay, average load, and number of collisions using the proposed protocol and standard TCP for the same network configuration. Fig. 6 shows the transmission delay observed at the network layer under the various load condition generated by periodic sources with 10 % of variance of transmission interval. The efficiency of PTM connection can be measured in this figure. There is significantly higher delay under conventional TCP than proposed protocol when

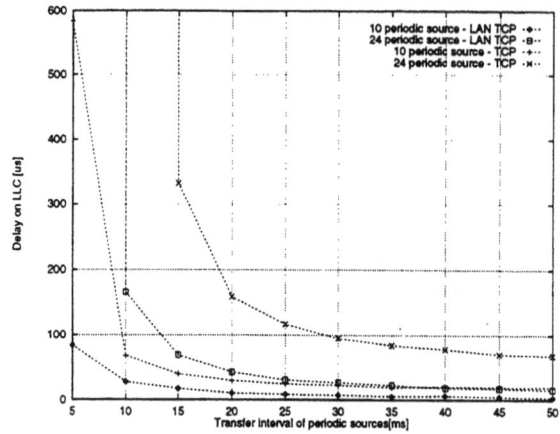

Fig. 6. Delay on LLC under loads generated by periodic sources (packet length is 258bytes)

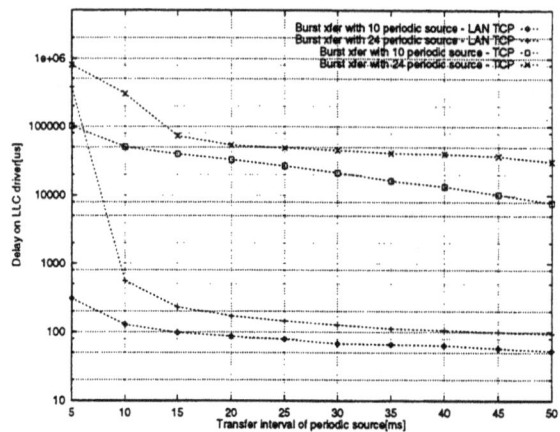

Fig. 7. Transmission delay caused by transient burst transmission

burst transmission is performed(Fig. 7). In a TCP burst transmission, average load can not exceed by unity because the transmission is based on packet exchange(Fig. 9). Even though this type of traffic is not continued long, it can cause critical problem in industrial network, especially very long delay, even transmission failure on MAC layer. From the simulation result, many transmission failures caused by 16 contiguous collisions are occurred as the number of collisions is increased (Fig. 8). The reason of huge delay is best-effort transmission which makes the average load around unity abruptly. Fig. 9 shows the effect of best-effort transmission under low load. For example, offered load is 348kbytes/sec with 25 mili-second of transfer interval by 24 TCP periodic sources. The load is increased to 1044kbytes/sec after burst transmission is started. This load results in significant transmission delay(Fig 7) and collisions(Fig 8). By the simulation result, 20~220 packets are dropped by transmission failure.

Of course, this number of collisions is not practical because burst transmission is never continued long during tens second. The number of collisions and transmission delay are varied by the start time of periodic transmission. Consider the

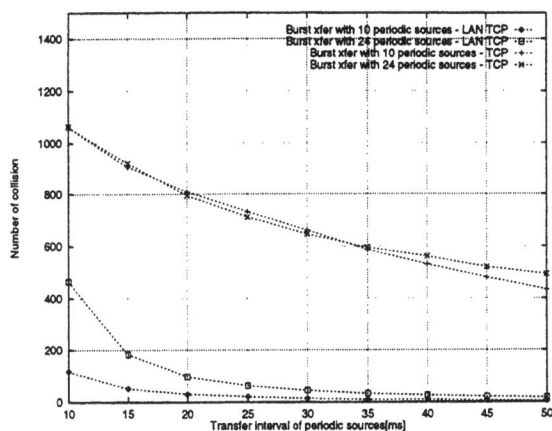

Fig. 8. Number of collisions under burst transmission

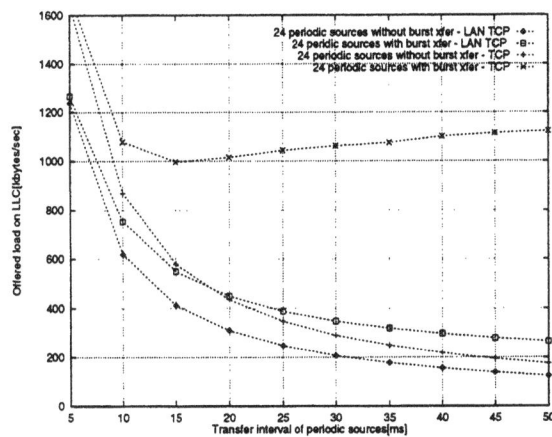

Fig. 9. A variation of channel utilization

worst case, if all nodes start periodic transmission and the transmission interval is not varied, then extreme transmission delay is occurred without burst transmission. On the other hand, if the start time is well distributed, then no collision is detected at all. So, periodic transmission which is based on system clock has potential problem.

If the simulation results(Fig 6, Fig 7, and Fig 8) are acceptable in real world, 20 periodic sources generating 200bytes of control data at each 10 mili-second and some LAN-TCP hosts can be feasible configuration.

5. CONCLUSION

This paper describes an effort to use industry standard TCP/IP - Ethernet for soft real-time systems including distributed computer control systems and manufacturing automation systems. To maintain the backward compatibility, the standard TCP header is extended to reduce traffic overhead and include periodic transmission mode. Because the proposed protocol extends the standard TCP/IP, all the existing application softwares the proposed TCP/IP can be used without modification. Besides soft real-time charac-

teristics, the proposed TCP/IP includes periodic transmission mode(PTM), which provides a very efficient transmission method of the periodical data for real-time monitoring and control systems. With PTM, periodical data collection and updating the control signal are possible with relatively small traffic overhead. Computer simulation shows the proposed TCP/IP has relative higher performance compare to the ordinary TCP/IP protocol.

The prototype system has been applied and evaluated to the large scale distributed control systems and human-like robot systems. Optimization of the proposed protocol is required to be used for a large scale distributed system and fine calibration may be necessary to maximize real-time performance to the base operating system.

6. REFERENCES

Boggs, D.R., J.C. Mogul and C.A. Kent (1988). Measured capacity of an ethernet: Myths and reality. *Proceeding of ACM SIGCOMM '88* pp. 222–234.

Chlamtac, Imrich, William R. Franta and K. Dan Levin (1979). BRAM: The broadcast recognizing access method. *IEEE Transactions on Communications* 27(8), 1183–1189.

Clark, David D., Van Jacobson, John Romkey and Howard Salwen (1989). An analysis of tcp processing overhead. *IEEE Communication Magazine* pp. 23–29.

Comer, Douglase (1995). *Internetworking with TCP/IP Volume I : Principles,Protocols, and Architecture*. Prentice Hall, Inc. Englewood Cliffs, New Jersey.

Comer, Douglase and David L. Stevens (1994). *Internetworking with TCP/IP Volume II : Design,Implementation, and Internals*. Prentice Hall, Inc. Englewood Cliffs, New Jersey.

ISI (1981). Transmission control protocol. *RFC793*.

ISI (1992). Extending TCP for transactions – concepts. *RFC1379*.

ISI (1994). T/TCP – TCP extensions for transactions functional specification. *RFC1644*.

Jacobson, V. (1990). 4BSD TCP header prediction. *Computer Communication Review*.

Lefelhocz, Christopher, Bryan Lyles, Scott Shenker and Lixia Zhang (1996). Congestion control for best-effort service: Why we need a new paradigm. *IEEE Network* pp. 10–19.

Metcalf, R.M. and D.R. Boggs (1976). Ethernet: Distributed packet switching for local computer networks. *Communication of the ACM* 19(7), 395–404.

Rom, Raphael and Moshe Sidi (1990). *Multiple Access Protocols*. Springer-Verlag. New York.

Shoch, John F. and Jon A. Hupp (1980). *Measured Performance of an Ethernet Local Network*. Xerox Corporation. California.

Signorile, Robert P. (1988). MBRAM-A priority protocol for PC based local area networks. *IEEE Network* **2**(4), 55–59.

Stevens, W. Richard (1994). *TCP/IP Illustrated, Volume 1*. Addison-Wesley Publishing Company. Massachusetts.

Tanenbaum, Andrew S. (1989). *Computer Networks*. Prentice Hall, Inc.. Englewood Cliffs, New Jersey.

Tobagi, F.A. and V.B. Hunt (1980). Performance analysis of carrier sense multiple access with collision detection. *Comput. Networks* **4**, 245–259.

Wright, Gary R. and W.Richard Stevens (1995). *TCP/IP Illustrated, Volume 2 : The Implementation*. Addison-Wesley Publishing Company. Massachusetts.

Yavatkar, Rajendra (1994). A reservation-based CSMA protocol for intergrated manufacturing networks. *IEEE Trans. on Systems,MAN,and Cybernetics* **24**(8), 1247–1258.

OPEN MODEL FOR THE
INTEGRATION OF VARIOUS FIELDBUS SYSTEMS

Tilo Pfeifer, Hong Seong Park°, Harald Thrum

Institute of Machine Tools and Production Engr.
Technical University of Aachen, 52056 Aachen, Germany
Tel. +49-241-80-7412, 7414
Fax. +49-241-8888-193
E-mail. (pf, thr)@wzl-mtq1.wzl.rwth-aachen.de

° Dept. of Control and Instrumentation Engr.
Kangwon National University
Hyoza 2-Dong, Chuncheon, Kangwon-Do
200-701, Korea
Tel. +82-361-50-6346
E-mail. hspark@cc.kangwon.ac.kr

Abstract: This paper deals with an object-oriented model for integration of different types of standardised and available fieldbus systems into one control system. This object-oriented model gives openness for the integration of various fieldbus systems, adds new functionalities and provides the user with an easy and common access method to sensors and actuators. As a selection of representative fieldbus systems PROFIBUS, CAN and INTERBUS are considered.

Keywords: Control applications, Control systems, Fieldbus, Sensors, Actuators, Object modelling techniques

1. INTRODUCTION

New concepts for the control of machine tools frequently require the increased integration of sensors and actuators. The collaborative research center "Autonomous Manufacturing Cells" (Weck, et al., 1996), funded by the Deutsche Forschungsgemeinschaft and involving several research facilities of the RWTH in Aachen, Germany, is a representative example of the use of numerous different sensors and actuators inside machine tools (Fig. 1).

An autonomous manufacturing cell of the type involved in this research, should be capable of performing complex machining processes reliably, without any malfunction, with a high level of precision and a maximum degree of autonomy over a longer period of time. It is envisaged that this

objective will be achieved by integrating concrete autonomous functions in the machine tool on-site. In particular, these will include functions relating to user interfaces, planning, control and monitoring as well as functions for workpiece and tool handling. The machining processes investigated in this instance are firstly, milling, which is a conventional machining technique commonly applied in industry and secondly, laser beam materials processing, a thermal technique which is used increasingly and is gaining in significance. These machining methods represent in different ways the fundamental problems associated with mastering manufacturing processes and demonstrate the requirement for action with regard to the implementation of suitable sensor/actuator systems and control techniques in order to increase machining accuracy, tolerance to faults and flexibility (Klocke, et al. 1995; Dahmen, et al. 1996).

Applications	switches	displacement/angle transducer	dynamometer, torquemeter	acceleration meter	temperature sensors	voltage, current and power meter	camera measurement systems	triangulation sensors	probes	microphones	sensors for acoustic emission	sensors for laser beam position, beam power, and intensity distribution	valves	drives	gripping/clamping systems	workpiece positioning	active spindle bearing
workpiece position and geometry registration							●	●									
process control with normal NC-, RC- and PLC functions	●	●	●		●	●							●	●			
user interface	●	●	●	●	●	●	●		●	●		●					
compensation of machine dislocation (static and dynamic)		●	●	●													●
laser beam monitoring						●						●					
tool wear and tear monitoring		●			●	●				●	●						
dimensional inspection							●	●	●								
workpiece handling	●		●			●							●	●	●	●	

Fig. 1. Sensors, Actuators and Applications for Autonomous Manufacturing Cells

2. REQUIREMENTS FOR THE INTEGRATION OF SENSORS AND ACTUATORS

Special sensors and actuators are integrated less in standard machines than in customized machines and in specialized manufacturing operations. In such cases, it is essential that the machine manufacturer is capable of reacting to individual customer requirements with purpose-built machine equipment. Sometimes it is the end user himself who wishes to complement or alter the machine. The more individual the requirements regarding the equipment are, the more frequent is the need to incorporate components made by a variety of manufacturers. But even when standard equipment is used, the option of being able to select manufacturers freely would intensify competition, as manufacturers vied with one another to produce better, lower-cost components.

The combination of products made by various manufacturers assumes, of course, that these are compatible with one another from a technical point of view. In terms of the sensors and actuators within machine tools, this is primarily a problem of information and communications technology and of electrical and mechanical device interfaces. The deployment of so-called "Open systems", whose technical details are specified by standards, transparent to both manufacturers and users, is necessary in such cases (Eversheim, et al., 1997).

In the field of data communication, open systems have not only been established for some time, but they have also made their mark clearly on the principle of openness in research and industrial practice. Open data communication was developed initially in the seventies and eighties mainly for data exchange between central computer systems. The requirement for open communication systems for linking sensors, actuators, in/output modules,

programmable controllers etc., did not emerge until later, along with the increasing decentralization of information technology. Corresponding standards for field and sensor /actuator bus systems followed soon after. Whereas hardly any standards existed for this lower network levels in the late eighties, numerous systems have, in the mean time, been standardized to meet various requirement profiles and have been implemented, where appropriate, in products for industrial control.

Each of the field and sensor/actuator bus systems which has held its own in the market place is predestined for a particular application. There is no one system for all applications. The leading standardized field and sensor/actuator bus systems in the field of production and manufacturing technology in the European market are PROFIBUS (DIN 19245, 1991), FIP (Solvie, 1993), INTERBUS (DIN 19258, 1994) and CAN (ISO-IS 11898). Internationally work is currently under way within the IEC (International Electronical Commission) on a world-wide field bus standard (IEC1158, 1992). Today, however, a standard exists only for the physical layer; draft standards exist for the data link and applications layers. It seems likely that European and American standards will have established themselves by the time a complete IEC standard, which takes account of a number of corporate interests as well as of technical and organisational issues, is passed. The specifications of PROFIBUS (Germany), FIP (France) and P-NET (Denmark) already exist as European Standard EN 50170 (Bötcher, 1996). Preparations are being made for the adoption of PROFIBUS as the an American ANSI standard as for that of INTERBUS as an European standard.

This clearly demonstrates that industrial users and manufacturers of automation components will have to cope with a variety of field and sensor actuator bus systems and their different, sometimes very complex interfaces and rules of application for some time to come.

If machine manufacturers and equipment companies are to be able to choose from the variety of sensors and actuators built by various manufacturers, they have to support several different interface standards. In the case of machine tools, this applies particularly to the control unit, since this as the central information processing component, must incorporate several decentrally recorded and pre-processed sensor signals and control the manufacturing process via several actuators.

In view of the ready availability of field and sensor/actuator bus systems, the problem of pure data communication as set out in the ISO-/OSI reference model, can be regarded as solved in terms of interfaces for sensors and actuators. The problem, however, remains of varying rules of application and access interfaces when several of these communications standards have to be used in order to ensure that the components need not be selected on the basis of device interfaces but can be selected for their functional performance.

Considerable efforts are now under way in research and industry to unify the application of various field and sensor/actuator bus systems (e.g.: Stiefenher,

1996; Pfeifer and Fussel, 1995). Also the research work presented by this paper makes a contribution to simplify the handling of different bus systems. But it focus mainly on the use of object-oriented techniques as well as the use of open and modular control units similar to the OSACA architecture (Pritschow, et al., 1993).

3. OBJECT-ORIENTED CONCEPT FOR THE INTEGRATION

The concept for integration of sensors and actuators via different interface standards by means of objected-oriented modelling technique is outlined in Fig. 2.

Via a communications system within a modular system control unit, a sensor/actuator module (in the following referred to as S/A module) supporting various bus systems can provide other modules with standardised interfaces and functions for access to sensors and actuators and for configuration and monitoring, independent of the underlying interface. So the application program itself requires only a minimum of information in order to access a sensor or actuator value e.g. for measurement or control. All functions relevant to communications and monitoring which were only "overhead" in relation to the applications within the control system are contained

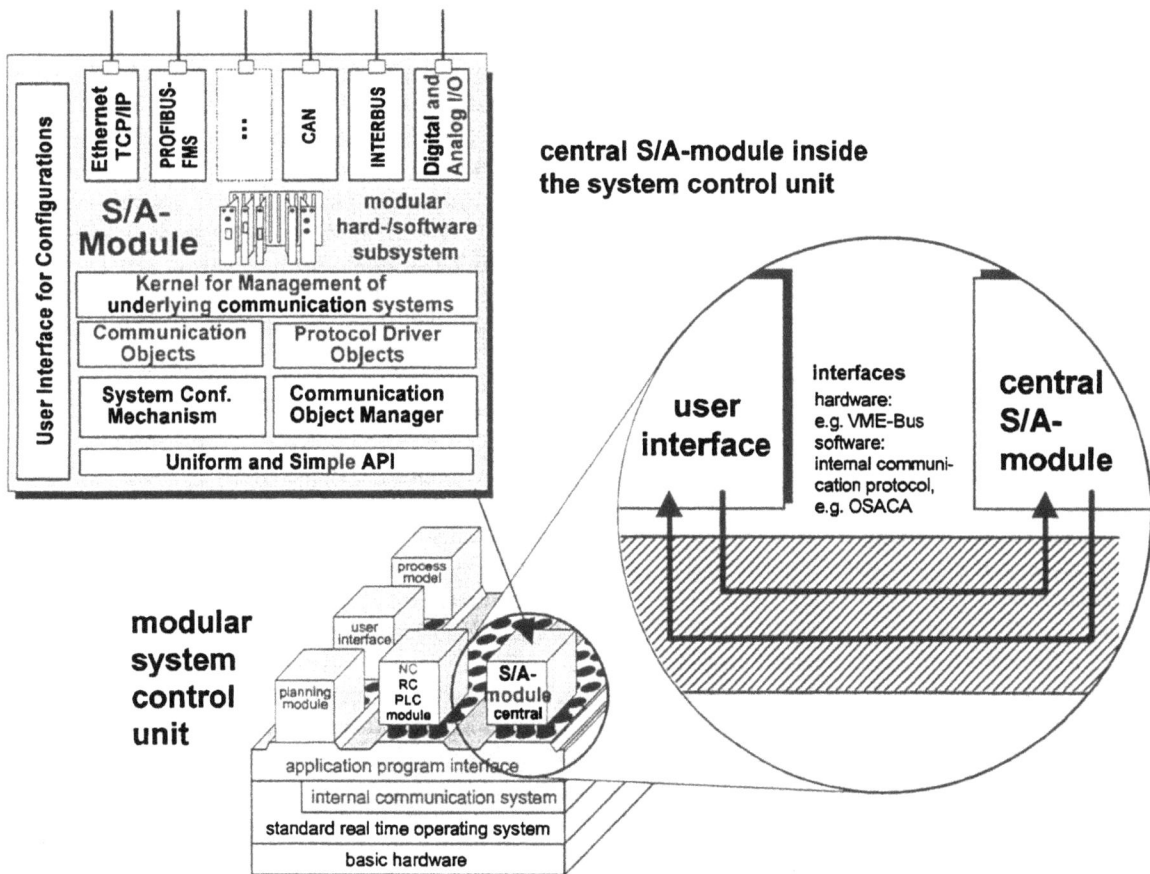

Fig. 2. Concept for Integration of the Sensors and Actuators into a Modular System Control

completely within the central S/A module and are no longer visible behind the interface in the control system.

The proposed concept can simplify considerably the programming, commissioning and monitoring of sensors and actuators within complex machining facilities such as the autonomous manufacturing cells described. This is mainly done by the features supported by the central S/A module. The development of this concept by means of the object-oriented modelling technique is described in the following section.

4. OBJECT CLASSES AND INTERFACES OF THE SENSOR/ACTUATOR MODULE

The concept for integration of sensors and actuators considers different types of open and commercial fieldbus systems. At the present state of realisation the bus systems PROFIBUS, INTERBUS and CAN are put together for monitoring and control applications with real time constraints.

Fieldbus systems can be classified into client/server and publisher/subscriber with regard to applications, multiple masters/slaves and one master/slaves with regard to bus access, and message oriented addressing, node oriented addressing, and hardware dependent addressing with regard to source and destination of data. In the following the fieldbus systems PROFIBUS-FMS (client/server, multiple masters/multiple slaves, node oriented addressing), CAN (publisher/subscriber, multiple masters, and message oriented addressing) and INTERBUS (client/server, one master/multiple slaves, and hardware dependent addressing) are described briefly.

The European standard PROFIBUS consists of PROFIBUS-FMS and -DP. PROFIBUS-FMS is a client/server model with multiple masters and multiple slaves, while PROFIBUS-DP is a client/server model with one master and multiple slaves like INTERBUS. This paper considers PROFIBUS-FMS for the integration of high functional sensor and actuator systems.

The international standard CAN adopts a publisher/subscriber model, where the publisher is a node transmitting messages to the bus and the subscriber is a node receiving messages from the bus. By that CAN supports multiple master systems. All nodes on the bus receive all transmitted messages and perform an acceptance test on the identifier to determine whether or not the message is relevant. If one received message is filtered as relevant, it will be available for further evaluation within the node.

The draft german standard INTERBUS is a client/server model with one master, typically inside one controller, and multiple slaves, typically I/O modules. The communication protocol is based on

ring topology, around shifting data and clock synchronisation. INTERBUS needs two cycles or two types of frames: identification (ID) cycle and scan cycle. During initialisation the scan cycle tells the master the type and position of the slave modules within the ring. During each scan cycle data is transferred from the master to output slaves and from input slaves to the master.

These three fieldbus systems together cover by their different properties a large spectrum of applications. But their diversity makes it difficult to use them together, especially under real time constraints.

From a point of view of both the application and the communication protocol the object-oriented model will be designed so that users can easily add and change the application program and the communication protocol. It provides two basic objects which are called a communication object (CO) class and a protocol driver (ProDrv) object class. The former manages objects relevant to variable domain, and event. The latter mainly deals with service primitives relevant to access to remote communication objects. In addition of those objects another type of objects for applications is used, which are called application-dependent objects (AO). The application-dependent object consists of some communication objects and some special functions for easy application. One example of application-dependent objects is an object with a PID control function, where communication objects correspond to sensors/actuators on the fieldbus systems and ProDrv object comprises service primitives supported by the underlying fieldbus system. The relationship between those object classes is shown in Fig. 3 and Fig. 4. Especially, the concept of AO is shown in Fig. 3. The AO has one or more COs so that its special application function is executed using these COs. CO has an association as reference to ProDrv object instance because CO should use the underlying fieldbus system to access to remote sensors/actuators.

4.1 Communication Object Class

The communication object (CO) is managed information about where to find which information and how to access it. COs considered in this paper are ones for variables, domains, events, and program invocations, which are denoted by variableCO, domainCO, eventCO, and piCO, respectively.

The class baseCO, which is the base for all communication object classes mentioned above, contains all attributes that are available in any communication object. It is an abstract class without any instances. The CO classes variableCO, domainCO, eventCO and piCO inherit services and attributes from baseCO.

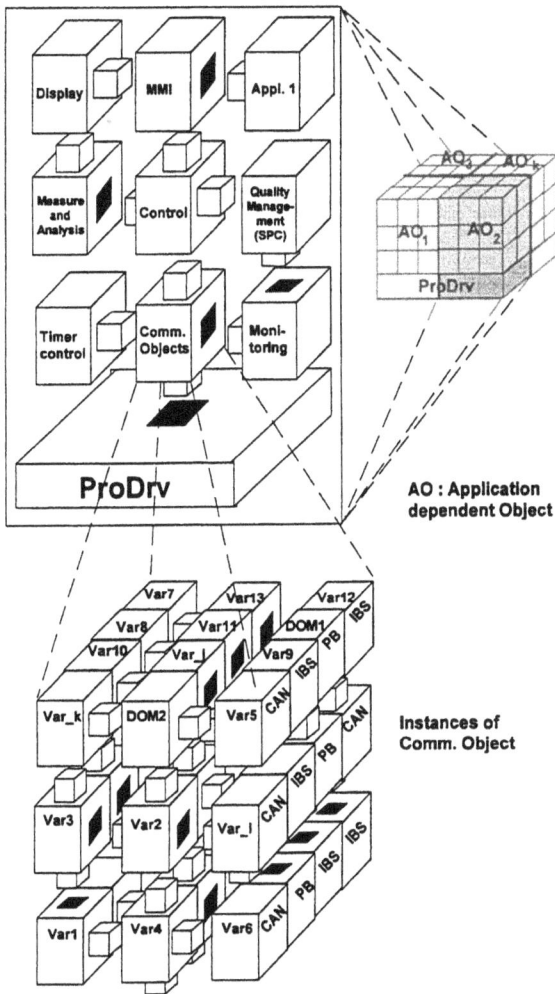

Fig. 3. Relationship between AO instances and other object instances

The communication object class variableCO serves for access to data. The data that is accessible by a variableCO for reading or writing can be located at a certain position in memory.

The CO class piCO serves for controlling process, task or programs. The execution of a piCO can be started, stopped, resetted and etc. The piCO class is a state machine with fixed sets of states, transitions and actions to trigger transitions. State transitions can occur through internal events.

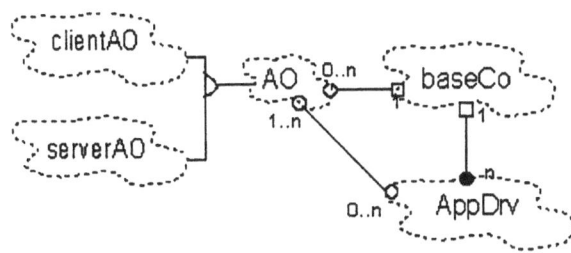

Fig. 4. Detailed Relationship between AO class, CO class and ProDrv class

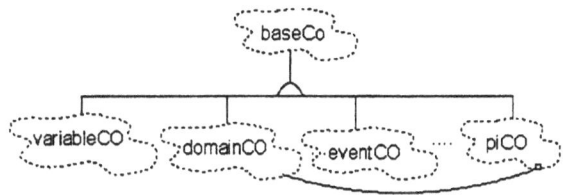

Fig. 5. Communication Object Classes

The eventCO is to report state changes in the equipment or in the logic of execution. Examples of events are alarm errors, execution cycles state changes and variable execution dependent changes. When a change in the value of an attribute is equal to one of given conditions, an automatic report to the client will be sent. The condition will contain one or more attribute, and values of instance.

The object class domainCO is used to manipulate files of data. The client AO can manipulate data files or memory data associated to a server. Typical examples of domains are machine programs, user defined programs or data, parameter tables, tool tables, system log data and configuration data.

As mentioned above, the Generalization-Specialization structure establishes inheritance realationship because CO object instance is able to have different object classes according to its usage. These relationship is shown Fig. 5.

4.2 Protocol Driver Object Class

The protocol driver (ProDrv) object class is responsible for the transport of message issued by AO and based on connection oriented type. So a connection must to be build up between two objects before transmission of message. During connection establishment all necessary parameter negotiations are performed. The ProDrv object class hides the specific capabilities of different underlying fieldbus systems. It provides a uniform set of services which is vendor independent and can select the suitable fieldbus system for each CO instance. The number of services is limited to make the protocol easy to use.

The services of ProDrv are corresponding to services or combined services of underlying fieldbus systems. For examples, the service initiateReq() corresponds to initiate service in PROFIBUS-FMS while domDownloadReq and piReq() have to use several relevant service primitives to be operated correctly.

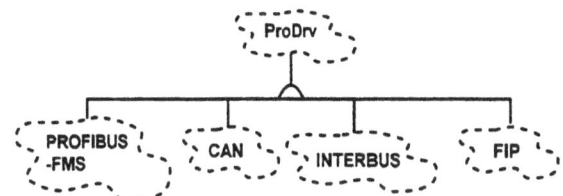

Fig. 6. Protocol Driver Object Classes

The ProDrv object class is a kind of abstract classes. So another object class for each fieldbus system must be defined, which is able to have a object instance. The Generalization-Specialization structure establishes inheritance relationship because ProDrv object instance is able to have the different object class according to underlying fieldbus systems. These relationship is shown Fig. 6.

5. CONCLUSIONS

It is obvious that the integration of sensors and actuators into control systems is still increasing and by that the usage of different fieldbus and sensor/actuator bus systems. In order to be able to exploit the full range of sensors and actuators available on the market and to integrate components made by various manufacturers it is necessary to take account of a number of interface standards like PROFIBUS, CAN and INTERBUS, which are leading for the European market.

The approaches presented here, demonstrate that the problem regarding data communication has already been solved by different standards and available products, but that there is still a requirement for action and a potential for optimisation with regard to the flexible usage of different standards inside one control system.

This paper presents an object oriented concept as a solution for the integration of different types of standardised and available fieldbus systems into one control system. The main part of this concept are two basic object classes: the communication object class and the protocol driver object class. The former manages objects relevant to variable, domain, event and program invocation. The latter manages objects with service primitives (or protocol) relevant to access of remote communication objects. Because these object classes provide a common basis for one fieldbus system or a combination of various ones, it gives users openness for the integration of fieldbus systems and for access to sensors and actuators.

Although the concept presented in this paper is based on an open architecture for machine tool control and considers only a few standardised bus systems (PROFIBUS-FMS, CAN and INTERBUS), it is also transferable to other modular control systems and interface standards. The concept presented in this paper can simplify considerably the programming, commissioning and monitoring of sensors and actuators within complex machining facilities such as autonomous manufacturing cells. The research work will continue with optimisations and tests for response time and synchronisation mechanism.

REFERENCES

Bötcher, J. (1996). EN 50170 - Die europäische Feldbusnorm - Entstehung, Spezifikationen und Folgen für Hersteller und Anwender, In: *Elektronik*, **12/96**, 11. June '96, Franzis, München

Dahmen, M., S. Kaierle, J. Kittel, and E.W. Kreutz (1996). Autonomous Manufacturing: Planning and Control in Laser Beam Welding. *Proceedings of the 15th International Congress on Applications of Lasers an Electro-Optics*, Detroit, Michigan, USA, 14.-17.10.1996

DIN 19245 German national standard. PROFIBUS, Process Field Bus, (1991) part 1 and 2, (1993) part 3 draft, (1995) part 4 draft

DIN 19258 German draft national standard (1994). INTERBUS, Sensor-/Aktornetzwerk für industrielle Steuerungssysteme

Eversheim, W., F. Klocke, T. Pfeifer and M. Weck (1997): Open Systems in Manufacturing, In: *Manufacturing Excellence in Global Markets*, 1997, p. 135-167, Chapman & Hall, London

IEC1158, part 2...7 draft, (1992...1995). Field bus standard for use in industrial control systems

ISO-IS 11898 (1993), Road vehicles - Interchange of digital information - Controller Area Network (CAN) for high speed communication

Klocke, F., W. König, G. Ketteler and M. Rehse (1995). Monitoring of Machining Processes using Intelligent Evaluation Strategies, *Proceedings of the Monitoring Machining Processes Clinic*, 12./13.09.1995, Detroit, USA, SME Technical Paper MR95-267

Pritschow, G., Ch. Daniel, G. Junghans and W. Sperling (1993). Open System Controllers - A Challenge for the Future of the Machine Tool Industry, *CIRP Annals 1993 Manufacturing Technology*, **Vol. 42/1/1993**, p. 449 ff., Verlag Technische Rundschau, Bern, Stuttgart

Pfeifer, T. and B. Fussel (1995). Integration von Feldbussen in die IEC 1131, Feldbustechnologie, Wien 26.-27.09.95, In: *Schriftenreihe des österreichischen Verbands für Elektrotechnik*, **no. 9**

Solvie, M. (1993). FIP - die französische Lösung, *pa – Produktionsautomatisierung*, **1/93**

Stiefenher, P. (1996). Macht NOAH das Feldbusschiff flott ?, *Markt&Technik*, **no. 18**, p. 32-34

Weck, M., N. Brouer, U. Herbst, F. Michels and D. Prust (1996). Sichere Prozeßbeherrschung und höhere Werkstückqualität durch autonome Produktionszellen, *wt - Produktion und Management*, **Vol. 86**, Springer, Berlin

SCHEDULING REAL-TIME MESSAGES IN FIBRE CHANNEL ARBITRATED LOOP

Jaeyong Koh, Taewoong Kim, and Heonshik Shin

Department of Computer Engineering
Seoul National University
Seoul 151-742, Korea
{jyk,twkim,shinhs}@comp.snu.ac.kr
Phone: +82-2-880-7295
Fax: +82-2-874-3104

Abstract: This paper addresses the problem of delivering real-time messages, both periodic and aperiodic, over fibre channel arbitrated loop. Periodic messages typically have hard real-time constraints whereas aperiodic tasks are delivered based on the best-effort approach. The deadlines of periodic messages are guaranteed by limiting the packet size under the access fairness algorithm defined in fibre channel arbitrated loop or FC-AL. The aperiodic messages are delivered using a global server scheme based on the full duplex communication and transfer state facilitated in FC-AL. The algorithm presented in this paper is optimal in that it guarantees all the deadlines of periodic messages and fully-utilizes network bandwidth. The optimality is expected to apply to many other broadcasting networks, both LANs and field-buses, by considering the global server scheme in their ongoing standardization processes.

Keywords: real-time communication, deadlines, scheduling algorithms, distributed computer control systems, communication protocols, local area networks

1. INTRODUCTION

Typical distributed control environment consists of a number of sampling devices, actuators, and administrative workstations that are connected by a network. This paper considers the problem of scheduling real-time messages in a distributed environment connected by Fibre Channel Arbitrated Loop or FC-AL (Sachs and Varma, 1996),(ANSI, 1996). FC-AL is a standard proposed by ANSI based on loop topology. Although it has been mainly exploited in SCSI disk interconnections, FC-AL is attractive for distributed applications covering up to several kilometers.

The sensors, actuators, and workstations connected to the network exchange periodic and aperiodic messages. Periodic messages are exchanged at regular intervals to monitor and control the system state. Aperiodic messages are generated at any time by external events such as user commands from the administrative workstation. The periodic messages must be delivered before their deadlines to ensure that the critical system state variables are maintained timely. These deadlines are called *hard* deadlines. On the other hand, the deadlines of aperiodic messages are not as stringent as periodic messages. It is desirable that aperiodic messages are transmitted as soon as possible without causing any periodic message to miss its deadline. That is, the aperiodic deadlines are considered *soft*.

The scheduling algorithm proposed in this paper guarantees the deadline of hard periodic messages and delivers soft aperiodic messages as fast as possible. The hard deadlines of periodic messages are guaranteed by limiting the transmission length per network access under the access fairness algorithm defined in FC-AL. The aperiodic messages are delivered whenever there is available time on the network using a global server scheme.

The rest of the paper is organized as follows. Problem definition, related works, and operation of FC-AL are described in Section 2. Section 3 deals with scheduling mechanism of periodic and aperiodic messages. Section 4 analyzes the algorithm proposed in this paper. Finally, Section 5 concludes this paper.

2. PROBLEM DEFINITION AND RELATED WORKS

2.1 *Real-Time Messages*

The system under consideration consists of multiple computing *nodes* connected by a broadcasting network. In the broadcasting network, one node, called the *sender*, gains access to the network and sends a message to another node, called the *receiver*, at one time.

Some messages are generated periodically and must be delivered before their deadlines, for example, sensory input and feedback control messages in chemical plants. Let *message stream* $S_i = (P_i, C_i)$ denote the sequence of periodic messages produced with a period of P_i and the maximum length C_i. Each message's deadline is assumed to be the end of its period. That is, a periodic message generated at time t must be delivered before time $t + P_i$. Multiple message streams may share the medium and are denoted as a set $S = \{S_i\}, i = 1, 2, \ldots, n$. Let $u_i = C_i/P_i$, representing per-stream utilization, and $u = \sum_{i=1}^{n} u_i$, the total utilization of periodic messages.

Aperiodic messages are generated at any time by external stimuli such as user commands. Let AP_i be the ith aperiodic message, $i = 1, 2, \ldots$. Each aperiodic message AP_i is described by its sender s_i, receiver r_i, and length l_i, that is, $AP_i = (s_i, r_i, l_i)$. Even though the exact timing characteristics of the aperiodic messages are determined only run-time, the average values of the timing characteristics are considered for network resource allocation. Especially, the average utilization of aperiodic messages, denoted u_{av}, is defined as

$$u_{av} = \frac{\text{avg.}(lengths\ of\ aperiodic\ messages)}{\text{avg.}(inter\text{-}arrival\ time)}.$$

The average utilization will be used as the criterion for allocating network resources for aperiodic communications.

2.2 *Related Works*

Intensive studies have been conducted for the transmission of real-time messages in LAN environments that consist of broadcasting networks with bus or ring topologies. In bus topology, VT-CSMA and window protocol have been proposed for Ethernet with CSMA-CD (Zhao and Ramamritham, 1987),(Zhao *et al.*, 1987). These schemes resolve the collisions of real-time messages based on the laxity of messages. However, since collisions cannot be avoided in CSMA-CD, these schemes cannot guarantee deadline of messages.

In ring topology, the collisions are completely avoided by rotating a token over the network. There are two representative token passing protocols in real-time communication. One is priority driven token ring and the other, timed token protocol such as FDDI.

Priority driven token ring supports a decentralized contention resolution based on the priority of message(Strosnider *et al.*, 1988). In this scheme, the token has priority and reservation fields. Each node attempting to access the network writes its priority on the reservation field of the token. To allow any higher priority pending message to reserve the network, a node must wait an additional rotation of the token before it actually sends its message. In addition, a node can send only one packet per network access so as to allow the fast preemption of the current communication by a higher priority pending message. Reservation based on priority and single packet per token enable deadlines of periodic message to be guaranteed. Transmission of aperiodic messages exploits Deferrable Server(DS) algorithm to enhance the communication performance (Lehoczky *et al.*, 1987). However, fixed priority assignment on packet results in low network utilization. Moreover, the scheme is not appropriate for high speed networks because it requires one additional token rotation before the actual data transmission and because it sends only a single packet per network access.

Unlike priority driven token protocol, the timed token protocol in FDDI bounds each node's media access time when it receives the token. Each node is assigned the synchronous bandwidth and it transmits synchronous(or periodic) messages up to the bandwidth. Each node knows a target token rotation time(TTRT) which denotes expected token rotation time. A node can transmit asynchronous (or aperiodic) messages only if the token has rotated the ring earlier than TTRT. When a token arrives late, a node transmits only synchronous messages.

The performance of FDDI depends on protocol parameters such as TTRT and synchronous bandwidths. Thus, previous works focus on the synchronous bandwidth allocation to meet the timing constraints of messages. (Agrawal *et al.*, 1992)(Chen *et al.*, 1992)(Agrawal *et al.*, 1993). (Chen *et al.*, 1992) proposed optimal allocation scheme. However, depending on the periodic message streams, the network utilization may drop below 40%.

2.3 *Overview of FC-AL*

FC-AL is a serial data channel that provides a logical bidirectional link between two *L_Ports*. Every node in FC-AL is connected to the network via L_Port as shown in Fig. 1. When a node attempts to access the network, it must win an arbitration. The arbitration relies on priority of each node. The access priority assigned to each node is based on its physical address in the network. The priority based arbitration can lead to situations where the lower priority L_Port cannot access the network. To prevent the lower priority node from starvation, FC-AL standard proposes *access fairness* algorithm.

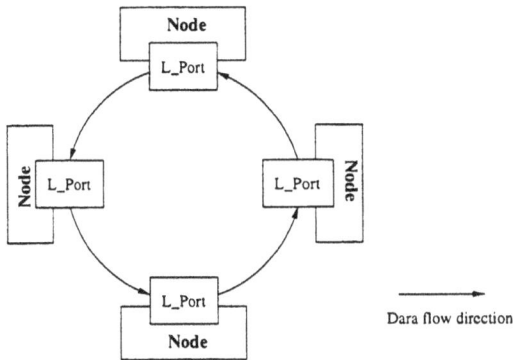

Fig. 1. *FC-AL network configuration*

Under the access fairness algorithm, if a node had won an arbitration and accessed the network, the node shall not request for arbitration again until the start of next *access window* where an access window is defined to be the time interval in which all the arbitrating L_Ports access the network. When the network becomes idle, a new access window starts.

Once a node wins the arbitration, it can open either half-duplex or full-duplex connection to the receiving node. See Fig. 2. Also, a node who won the arbitration can communicate with several destinations sequentially without additional arbitrations via *transfer state*. The full-duplex connection and the sequential transfer of connections to multiple receivers(called *transfer procedure* in the sequel) will play an important role in the real-time scheduling proposed in this paper.

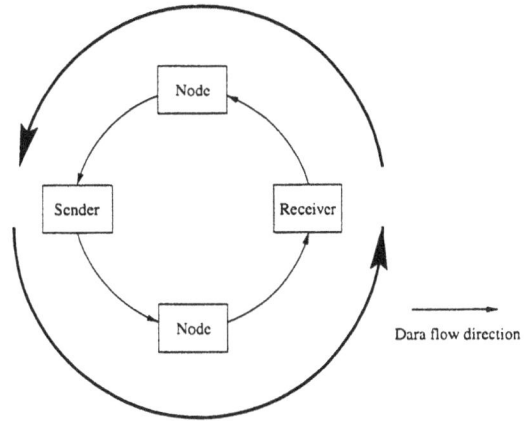

Fig. 2. *FC-AL full duplex communication*

3. DELIVERING REAL-TIME MESSAGES OVER FIBRE CHANNEL

3.1 *Scheduling Periodic Messages*

Even though the MAC provides fair access mechanism, FC-AL does not necessarily guarantee the transmission of hard real-time messages before their deadlines. According to access fairness algorithm, if a node has already accessed the ring once and intends to access again, it must wait for the next access window, delayed for the sum of transmission times contributed by other nodes. Consequently, to prevent any message from missing deadline, each node generating S_i messages should have the limit or *packet size*, denoted by f_i, on the duration it can transmit every time it wins the arbitration as shown in Fig. 3. The basic scheme introduced in this paper uses proper *packet size* f_i along with FC-AL's access fairness to guarantee deadlines of hard real-time periodic messages. It will be shown that the use of the proper packet sizes guarantees the real-time delivery of periodic messages without any additional operations such as clock synchronization.

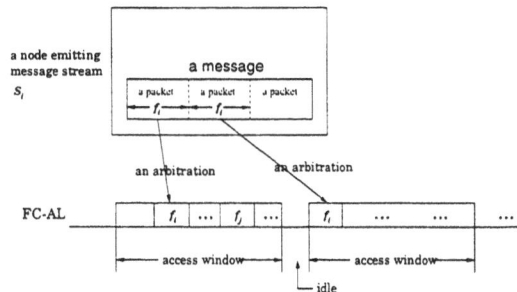

Fig. 3. *Access fairness algorithm*

The packet size f_i is obtained as follows: Let $A_i(t)$ denote the average amount of communication accomplished for S_i within time interval $[0, t]$. Under access fairness algorithm, each node emitting

S_i will have equal number of network accesses. Consequently, for two different message streams S_i and S_j, the relative amount of communication $A_i(t) : A_j(t)$ is identical to to relative packet size $f_i : f_j$. Generalizing this argument,

$$A_1(t) : A_2(t) : \ldots : A_n(t) = f_1 : f_2 : \ldots : f_n. \tag{1}$$

The messages that each stream S_i produces amount to $U_i t$ on average within the time interval $[0, t]$. Since all the produced messages should be transmitted, it follows that $A_1(t) : A_2(t) : \ldots : A_n(t) = u_1 : u_2 : \ldots : u_n$. From Eq. (1),

$$f_1 : \ldots : f_i : \ldots : f_n = u_1 : \ldots : u_i : \ldots : u_n.$$

The result can be summarized as

$$f_i = au_i \tag{2}$$

where a is a constant independent of $i = 1, \ldots, n$.

The width of any access window cannot exceed $\sum_{i=1}^{n} f_i = a \sum_{i=1}^{n} u_i = au$, so a sender of S_i will never be delayed for more than au time units each time it arbitrates to send a packet of size f_i. Within the time interval from the release time t to the deadline $t + P_i$ of a message in S_i, there are at least $\lfloor P_i/au \rfloor$ access windows in each of which S_i can send a packet. Consequently, the stream can transfer at least $f_i \lfloor P_i/au \rfloor$ within the time interval. Since the stream must transfer a message of size C_i within the time interval (and from Eq. (2)), it follows that

$$au_i \lfloor \frac{P_i}{au} \rfloor > C_i. \tag{3}$$

The parameter a must be identified which satisfies Eq. (3) for all $i = 1, \ldots, n$. Then, the packet sizes f_i, $i = 1, \ldots, n$ follow immediately from a. Let $Y(a) = au_i \lfloor P_i/au \rfloor$, the left side of the Constraint (3). The function $Y(a)$ is a piecewise linear function in terms of a. Let $a(k) = \frac{P_i}{ku}$ for all the positive integers $k = 1, 2, \ldots, \infty$. Within each interval $[a(k+1), a(k)]$ of the parameter a, $Y(a)$ becomes the linear function $Y(a) = ku_i \cdot a$ as shown in Fig. 4. Consequently, as illustrated in Fig. 4, the set of parameters a that satisfy the condition stated in Constraint (3) within the interval $[a(k+1), a(k)]$ are:

- $a \in [P_i/k, a(k)]$ if $P_i/(k+1)u < P_i/k$.
- The entire interval $[a(k+1), a(k)]$: $Y(a)$ is always bigger than C_i within the interval if $P_i/(k+1)u \geq P_i/k$ or, equivalently, $k \geq \lceil u/(1-u) \rceil$.

With A_i denoting the set of parameters a that satisfy Constraint (3) for some index i, $1 \leq i \leq n$, the above results can be summarized as

$$A_i = \{a \in [0, \frac{P_i}{\lceil u/(1-u) \rceil u}]\} \cup$$
$$\bigcup_{k=1}^{\lfloor u/(1-u) \rfloor} \{a \in [\frac{P_i}{k}, \frac{P_i}{ku}]\}. \tag{4}$$

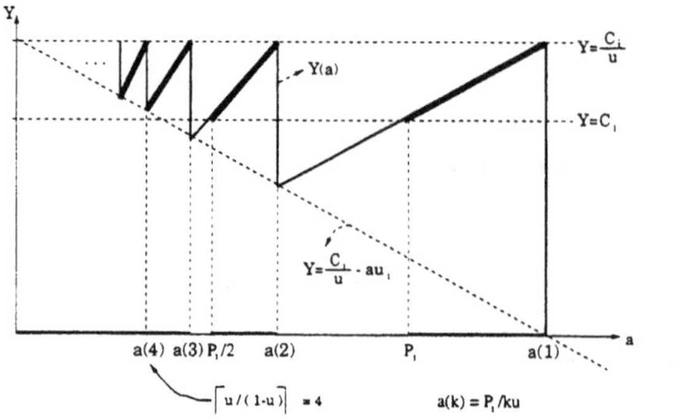

Fig. 4. *The feasible values of the parameter a*

To satisfy Constraint (3) for all message streams S_i, $1 \leq i \leq n$, the parameter a should belong to the set $A = \bigcap_{i=1}^{n} A_i$.

$$A = \bigcap_{i=1}^{n} \left[\{a \in [0, \frac{P_i}{\lceil u/(1-u) \rceil u}]\} \cup \bigcup_{k=1}^{\lfloor u/(1-u) \rfloor} \{a \in [\frac{P_i}{k}, \frac{P_i}{ku}]\} \right]. \tag{5}$$

Any packet size $f_i = au_i$ for $i = 1, \ldots, n$ where $a \in A$ will satisfy the deadline constraints of the set S of the message streams. The larger the value of a is, the less the protocol overhead becomes. Note that A never results in a null set because the value a ranging from gcd to gcd/u will always satisfy Constraint (3) for all $i = 1, \ldots, n$ where gcd is the greatest common divisor of all periods $P_1, \ldots, P_n$.

Example: Consider a set of message streams $\{S_1 = (6, 2), S_2 = (15, 3), S_3 = (30, 6)\}$. If messages are transmitted without being divided into packets, some of them may miss deadlines as shown in Fig. 5. The values of the parameter a which guarantee deadlines of messages in each S_i are obtained from Eq. (4).

$$A_1 = \{[0, 2.73], [3, 4.09], [6, 8.18]\},$$
$$A_2 = \{[0, 6.82], [7.5, 10.23], [15, 20.45]\},$$
$$A_3 = \{[0, 13.64], [15, 20.45], [30, 40.91]\}.$$

The values of a that guarantee deadline constraints of the entire set S are obtained as $A = A_1 \cap A_2 \cap A_3$. In this example,

$$A = \{[0, 2.73], [3, 4.09], [6, 6.82], [7.5, 8.18]\}.$$

As discussed above, the a's range $[gcd, gcd/u] = [3, 4.09]$ is included in the set A. However, to reduce the protocol overhead, the largest value of the set A is used, that is $a = 8.18$. It follows that the packet size for each stream is

$$f_1 = au_1 = 2.73,$$
$$f_2 = au_2 = 1.64,$$
$$f_3 = au_3 = 1.64.$$

The timeline of the message streams scheduled according to the algorithm presented in this paper is shown in Fig. 6. No message misses its deadline. □

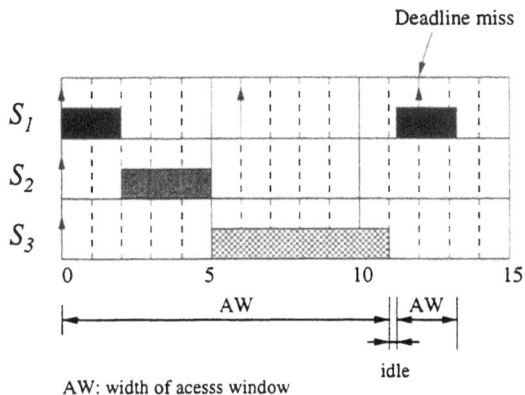

Fig. 5. *Problem of access fairness algorithm*

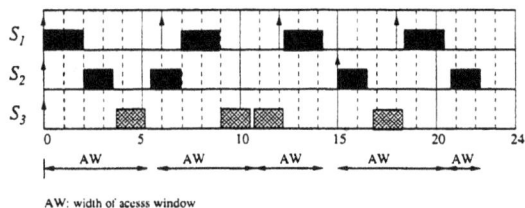

Fig. 6. *Timeline of periodic messages*

3.2 *Scheduling Aperiodic Messages*

Aperiodic messages are generated at any time and should be transmitted as fast as possible using the idle time of the network. More than one nodes may send aperiodic messages which are arbitrated by the access fairness algorithm together with the periodic messages. The total amount of aperiodic communication within each access window should be limited or some periodic messages may miss their deadlines. Let b be the upper bound on the total communication time of aperiodic messages within an access window. Let U_{ap} be the the network utilization allocatable to aperiodic messages. U_{ap} is the portion b among the entire access window whose width equals $au + b$, that is,

$$U_{ap} = \frac{b}{au + b}. \qquad (6)$$

To deliver all the aperiodic messages, U_{ap} must be larger than the average utilization u_{av} defined in Section 2.1. It is also obvious that U_{ap} cannot be greater than $1 - u$, the remaining utilization unused by periodic messages. Depending on system requirements, U_{ap} can be chosen to be the value that falls within the interval $[u_{av}, 1-u]$. If a larger value of U_{ap} is used, the system can allocate more utilization to aperiodic tasks, reducing the average response time of the aperiodic communication. However, the larger the U_{ap}, the smaller the packet size of periodic messages. This implies that the larger U_{ap} incurs the higher overhead for protocol operations.

Once the parameter U_{ap} is determined considering the above trade-off, the parameter b becomes a function of the parameter a, i.e., from Eq. (6),

$$b = \frac{U_{ap}}{1 - U_{ap}} au. \qquad (7)$$

To guarantee the delivery of each periodic message of length C_i before its deadline P_i in the existence of the aperiodic communication, the following constraint must be satisfied.

$$au_i \lfloor \frac{P_i}{au + b} \rfloor > C_i. \qquad (8)$$

From Eq. (7) and Constraint (8), it follows

$$au_i \lfloor \frac{P_i(1 - U_{ap})}{au} \rfloor > C_i. \qquad (9)$$

The proper parameter a is determined which satisfies the above constraint. Once the proper value of a is identified, the parameter b immediately follows from Eq. (7) The constraint (9) is identical to the one shown in (3) with P_i substituted by $P_i(1 - U_{ap})$. Consequently, the set A of feasible values of a is obtained similarly.

$$A = \bigcap_{i=1}^{n} \left[\{a \in [0, \frac{P_i(1 - U_{ap})}{\lceil u/(1-u) \rceil u}]\} \cup \bigcup_{k=1}^{\lfloor u/(1-u) \rfloor} \{a \in [\frac{P_i(1 - U_{ap})}{k}, \frac{P_i(1 - U_{ap})}{ku}]\} \right].$$

$$(10)$$

Eq. (10) shows that, as the amount of utilization U_{ap} allocated to aperiodic communication increases, the parameter a decreases incurring a large overhead for protocol operations for periodic communications as discussed previously.

However, the large U_{ap} implies the faster aperiodic communication or the swifter recovery from transient overload of aperiodic messages.

Eq. (10) and Eq. (7) derive the values of parameters a and b from the properties of periodic message streams and the aperiodic communication. When the system starts with the periodic message streams, the parameters are determined and broadcasted to all the nodes in the network. The nodes then divide their periodic messages into the packets of length au_i and the aperiodic messages into packets of length b. If a single node sends both periodic and aperiodic messages, each pair of the packets, periodic and aperiodic, cannot be sent separately within an access window due to the access fairness algorithm. In this case, the aperiodic packet can be piggy-backed to the periodic packet and both are sent within an access window if the receivers are identical. Otherwise, the two messages can be sent to the different receivers using the transfer procedure defined in FC-AL.

Multiple nodes may attempt to send aperiodic messages simultaneously. If all of them are allowed to access the network within an access window, the upper bound b may not be observed. To limit the total amount of aperiodic communication within an access window, transmissions of some messages must be delayed until the next or subsequent access windows. To enforce such delay over FC-AL network, a node is elected server that monitors network status and enforces the required delay for some of the simultaneous aperiodic messages. The server node is selected among ordinary nodes(hopefully, lightly loaded one) by a ring-based distributed algorithm whenever network configuration changes (Chang and Roberts, 1979).

The overview of the server scheme is described as follows:

Step 1: For the ith aperiodic message AP_i, its sender s_i first connects to the server(Fig. 7a).

Step 2: The server replies back the the amount of delay the node must wait for before the actual message transmission. If the network has not recently been used for aperiodic message transmission, the server returns zero delay(Fig. 7b). Otherwise, it returns non-zero delay to observe the bound of the aperiodic communication within each access window.

Step 3: If zero delay is replied from the server, the sender s_i transfers the connection from the server to its receiver r_i of the message and actually send the message(Fig. 7c).

Step 4: Otherwise, the sender waits for the delay and then attempts to send the message within next or subsequent access windows.

All these actions can be done in one indivisible step using the full-duplex connection and the transfer procedure defined in FC-AL. That is, the steps 1 and 2 are accomplished within single full duplex connection. Steps 2 and 3 are accomplished without additional network arbitration using the transfer procedure. These processes incur negligi-

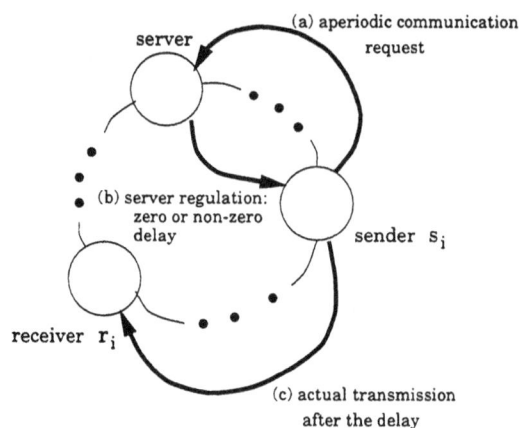

Fig. 7. *The global server scheme for transmission of aperiodic messages*

ble overhead. The delay calculated from the server clock can be used in the sender safely without clock synchronization since the delay is the length of time, not the actual time point.

The requests for transmission of the aperiodic messages $AP_1, AP_2, \ldots, AP_i, \ldots$ arrive at the server sequentially at time $t_1, t_2, \ldots, t_i, \ldots$ where $t_i < t_{i+1}$. If the number of pending aperiodic messages is more than b, the messages are divided into packets of length b, each of which is transmitted every access window. If the length l_i of message AP_i is larger than b, AP_i is divided into packets of size b and transmitted separately as shown in Fig. 8. To guarantee the deadlines of periodic messages even in their worst case, the server requires that the width of access window in which each packet is transmitted be $au + b$. When the next aperiodic message AP_{i+1} is requested while the transmission of AP_i is ongoing, the server calculates the delay for which s_{i+1} must wait before the transmission of AP_{i+1} based on the initiation time and length of previous message AP_i.

Since the length l_i of the message AP_i is not an integer multiple of b in general, the last packet of the message may be shorter than b. Within single access window, the last packet can be scheduled together with the first packet of the next message as shown in Fig. 8. Thus, the server also keeps track of the length of the last packet of message to determine the first packet size of the next message, which is transmitted together with the last one. Although there may be multiple pending aperiodic messages $AP_{i+1}, AP_{i+2}, \ldots, AP_j$, it is clear that only two variables need to be maintained to service the next arrival of aperiodic request, AP_{j+1}: the delay and the length of the last packet of AP_j.

3.3 Resource Reclaiming

The aperiodic transmission scheme proposed in Section 3.2 introduces delays before the actual

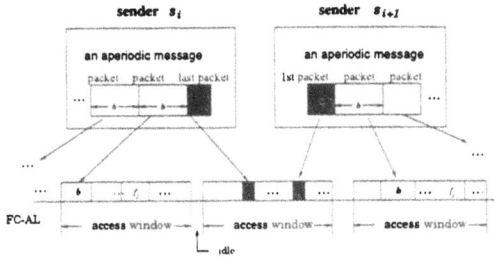

Fig. 8. *Transmission scheme of aperiodic messages: Packet size*

message transmissions. The delays, however, were calculated based on the assumption that the width of access window is uniformly $au + b$, that is, the lengths of all periodic messages are C_i, the maximum possible length of the actual messages. In many applications, the lengths of periodic messages may vary, i.e., they are typically smaller than C_i. Consequently, the senders of aperiodic messages may initiate their transmissions earlier by reclaiming the network bandwidth unused by the periodic messages.

Another aperiodic transmission scheme is proposed in this section to reclaim the unused bandwidth. In the new scheme, the sender s_i not only connects to server at the start of AP_i transmission, but also informs s_{i+1} of the completion of AP_i at the end of the transmission. The successive sender s_{i+1} immediately initiates its arbitration when it is informed by s_i, achieving the full utilization of the network. To this end, the server maintains *network-wide ready queue* or *NRQ* where the pending senders of aperiodic messages are queued for their turn as shown in Fig. 8. Only the *head* or the first node in NRQ transmits the aperiodic message. No delay calculation is required.

The detailed algorithm is described as follows: The server maintains a pointer to the *tail* or the last node in the NRQ. Each sender s_i points to its immediate successor s_{i+1} in NRQ. Assume the NRQ is populated by the senders $s_i, s_{i+1}, s_{i+2}, \ldots, s_j$. The server points to the tail, s_j. The node s_i currently occupies the network and continues transmitting the aperiodic message. Two types of events can happen which are processed as follows:

- **Completion of Message Transmission:** If the node s_i completes transmitting the message, it transfers its connection to the successor s_{i+1} and wakes up s_{i+1} for next aperiodic transmission. If the node s_i has no more successors, it informs the server that NRQ is now empty.
- **New Aperiodic Request:** When the new sender s_{j+1} requests the server for the aperiodic transmission, the server replies the pointer to the tail s_j through full duplex connection. s_{j+1} transfers its connection to

s_j to inform that the successor of s_j is s_{j+1} itself. It then sleeps in the queue until s_j wakes it up. The server records s_{j+1} as the new tail of NRQ. See Fig. 9. Note that, if NRQ is empty when s_{j+1} requests for the transmission, the server informs that s_j is the new head of NRQ and s_j directly sends first packet to the receiver. These step are done by full duplex connection and transfer procedures with negligible overhead.

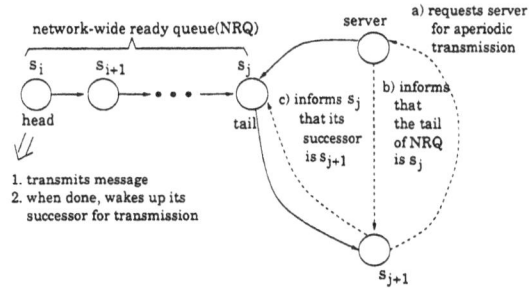

Fig. 9. *The network-wide ready queue scheme for aperiodic message transmission with resource reclaiming*

The new scheme for aperiodic communication reclaims the network resource unused by the periodic communication. The algorithm is simpler than the one introduced in Section 3.2. However, it requires more node processing, that is, the new scheme need to interrupt (1) the server, (2) the tail of NRQ, and (3) the successor in NRQ whereas the previous scheme interrupts the server once. Since existing FC-AL protocol does not provide the server scheme, the node processing must be implemented as an application program or kernel process. Although the processing requires very simple arithmetic or pointer operations, it needs CPU interrupt, context switching, and message processing overhead when implemented as a program. Thus, it is strongly recommended that the FC-AL protocol is extended to include the simple server scheme introduced in this section.

4. ANALYSIS OF THE ALGORITHM

The message transmission algorithm presented in this paper has some important properties. First, it guarantees all the deadlines of periodic messages for any set of message streams unless the network is physically overloaded($u \leq 1$). The proof on the optimality of the algorithm for periodic messages is as follows:

Theorem 1: (Optimality of the algorithm) *For any set S of message streams, the deadlines of their messages are satisfied unless the network is physically overloaded.*

Proof: For any message in S_i arriving at t, the amount T of transmission accomplished by the deadline $t + P_i$ is, (as in Eq. (9)

$$T = au_i \lfloor \frac{P_i(1 - U_{ap})}{au} \rfloor.$$

Substituting $gcd(1 - U_{ap})/u$ for a, we obtain

$$T = \frac{1 - U_{ap}}{u} \cdot gcd \cdot u_i \lfloor \frac{P_i}{gcd} \rfloor.$$

Since all P_i's, $i = 1, \ldots, n$, are the integer multiples of gcd,

$$T = \frac{1 - U_{ap}}{u} C_i.$$

If the system is not physically overloaded, $U_{ap} + u < 1$. Consequently, $T > C_i$ for all $i = 1, \ldots, n$ and thus all the messages are delivered within their deadlines. □

Another important property of the algorithm is that it fully utilizes the network for the transmission of soft aperiodic messages. Section 3.3 clearly showed that the algorithm transfers pending aperiodic messages whenever the network bandwidth is available.

5. CONCLUSION

This paper has discussed the problem of guaranteeing the deadlines of periodic messages and delivering aperiodic messages over FC-AL. The deadlines of periodic messages are guaranteed by limiting the transmission length per arbitration. The scheme is optimal in that it guarantees periodic messages of up to 100 % utilization. The aperiodic messages are delivered by the global server scheme together with the periodic messages. The proposed scheme does not incur the high overhead such as additional rotation of tokens as in the priority driven token protocol. Compared with FDDI which uses local timers, the proposed scheme achieves higher utilization of the network based on the global server.

In summary, the algorithm utilizes full bandwidth of the FC-AL while it guarantees respective deadlines of hard real-time messages. The results shown in this paper are expected to apply to many other broadcasting networks, both LANs and field-buses, by considering the simple global server scheme in their ongoing standardization processes.

6. REFERENCES

Agrawal, G., B. Chen and W. Zhao (1993). Local synchronous capacity allocation for guaranteeing message deadlines with the timed token protocol. In: *Proceedings of IEEE INFO-COMM*. pp. 186–193.

Agrawal, G., B. Chen, W. Zhao and S. Davari (1992). Guaranteeing synchronous message deadlines with the timed token protocol. In: *Proceedings of IEEE Conference on Distributed Computing Systems*. pp. 468–475.

ANSI (1996). Fibre channel arbitrated loop. Draft proposed-ANSI Standard X3T11, rev 5.2.

Chang, E. G. and R. Roberts (1979). An improved algorithm for decentralized extreme-finding in circular configuration of processors. *Communication of the ACM* **22**(5), 281–283.

Chen, B., G. Agrawal and W. Zhao (1992). Optimal synchronous capacity allocation for hard real-time communications with the timed token protocol. In: *Proceedings of Real-Time Systems Symposium*. pp. 198–207.

Lehoczky, J. P., L. Sha and J. K. Strosnider (1987). Enhanced aperiodic responsiveness in hard real-time environment. In: *Proceedings of the Real-Time Systems Symposium*. pp. 261–270.

Sachs, M. W. and A. Varma (1996). Fibre channel and related standards. *IEEE Communications Magazine* **34**(8), 40–50.

Strosnider, J. K., T. Marchok and J. P. Lehoczky (1988). Advanced real-time scheduling using the ieee 802.5 token ring. In: *Proceedings of the Real-Time Systems Symposium*. pp. 42–52.

Zhao, W. and K. Ramamritham (1987). Virtual time csma protocols for hard real-time communication. *IEEE Trans. Software Engineering* **13**(8), 938–952.

Zhao, W., J. A. Stankovic and K. Ramamritham (1987). A window protocol for transmission of time constrained messages. *IEEE Trans. Computers* **39**(9), 1186–1203.

118

TRENDS IN INTEGRATED SHIP CONTROL NETWORKING.

Niels Jørgensen * Jens F. Dalsgaard Nielsen *

* Aalborg University
Department of Control Engineering
Fredrik Bajersvej 7C, 9220 Aalborg E, Denmark
mailto: {jdn,nj}@control.auc.dk, http://www.control.auc.dk

Abstract: Integrated Ship Control systems can be designed as robust, distributed, autonomous control systems. The EU funded ATOMOS and ATOMOS II projects involves both technical and non technical aspects of this process. A reference modeling concept giving an outline of a generic ISC system covering the network and the equipment connected to it, a framework for verification of network functionality and performance by simulation and a general distribution platform for ISC systems, The ATOMOS Network, are results of this work.

Keywords: Distributed Computer Control Systems, Network Reliability, Real-time Communication

1. INTRODUCTION

As a new trend, Integrated Ship Control (ISC) suggests an architechture which integrates all control tasks on board a ship into a single concept (Dittmann, 1992). From this point of view an ISC system can be designed as a robust, distributed, autonomous control system. The EU funded ATOMOS and ATOMOS II projects involves both technical and non technical aspects of this process.

An ISC system is by nature (and history) a hetorogenic system, utilizing equipment from different vendors. In the ATOMOS project much effort is put into modeling a generic ISC system covering aspects from human factors to pure technical details.

Our part of the project covers mechanisms for integration of different subsystems such as main engine, power management, maneuvering (rudder, thrusters), navigational equipment, etc. into a homogeneous system.

This paper presents following results from this work:

- A reference modelling concept giving an outline of a generic ISC system covering the network and the equipment connected to it. For the purpose of verifying the network system, traffic information must be specified as a part of the model.
- The model enables a verification of the network functionality and performance by simulation. The framework and goals for the simulation are presented briefly.
- To serve as the general distribution platform in ISC systems the ATOMOS network is designed and prototyped. The key features of the ATOMOS network is presented.

2. THE ATOMOS NETWORK

The network serves as the environment for integrating the different ISC subsystems into one distributed control system providing high quality real-time services as the main goal (Nielsen et al., 1995).

Due to the real-time nature of the traffic, extensive non real-time traffic (like file transport) is

119

restricted. A gateway functionality to non real-time networks is thus provided.

Classification societies, as 'Lloyds' and 'Norsk Veritas' defines strict demands on functionality, timing and robustness for ISC systems. Following key demands exists for ISC networking:

- Ability to mask at least one failing component.
- Ability to support a one second maximum end to end alarm notification time.
- Deterministic response time, allowing time critical communication as syncronisation and closed loop control.
- Graceful degradation in case of faults and overloads including traffic priorities.
- Dynamic connection and disconnection of nodes with a minimum of interruption.

In collaboraton with our industrial partners following demands on capacity and timing has been evolved:

- Recovery time after critical communication error (ex. token loss) max. 200 ms.
- Total network segment capacity on at least 2000 messages/second.
- At least 30 nodes allowed on each network segment,
- At least 1000 m segment length. (Non intelligent repeaters allowed.)
- Medium access time less than 5 msec. on a network segment with 10 nodes.
- Low price of network controllers (ie. max. 80 US$).

No general demands exist on capacity of the individual network nodes, but at least 200 messages/second are suggested.

The listed figures are an estimation of communication needs in a medium to large cargo vehicle with a high degree of automation and integration. Large passenger and cruising vehicles is estimated to have higher demands on traffic. To meet higher traffic demands a segmentation of the network can be necessary.

2.1 Selection of Standards

Available standards for control networks and field busses were extensively examined (Jensen and Granum, 1993) (Hansen and Granum-Jensen, 1993) to find either a complete solution, or components fitting our needs. In the examination the above listed demands were used together with the following design criterias:

- To maintain the view of ISC systems as true distributed systems, it was decided at an early stage to disqualify pure master-slave structures in the communication system.

- Commercially available controllers or interface components should exist.

Following standards was examined: IEEE-802.x, Proway C, ARCnet, Cambridge Ring, Lon Works, MAP, MiniMAP, FMS, MIL-STD 1553B, IEC/ISA SP50, Profibus, InterOperable Systems Fieldbus and CAN Bus. The fact that the standards differs a lot (some covers all OSI layers, some just a few) makes direct comparisons difficult. For standards including OSI layer 1 and 2 figures comparable to the list of capacity, timing and price demands was calculated. For standards including application interfaces, characteristics of these was recorded.

Following selections was made:

- For OSI layer 1 and 2 ARCnet was selected. ARCnet uses a simple token bus protocol, it offers up to 2.5 Mbit bandwidth and cheap LSI controllers are available.
- As application interface FMS was selected. FMS is derived from MAP and is used in some fieldbusses (as Profibus).

2.2 Design

To enable single fault masking a dual network approach using dual network controllers was selected. All network functionality was made independent of host loads and timing by imploying a local processor for protocol handling. To support this structure, software covering OSI layer 3-6 was developed (Granum et al, 1993).

It was found that having two separate networks, although more expensive would give a more flexible solution than just using double cabling. Bearing in mind that the network is designed to exist in heterogenious environment it was decided to make all network functionality independent of host loads and timing by imploying a local processor for protocol handling. The system was also designed to allow uncritical, cheap equipment using non redundant connections. Figure 2 shows the structure of network and network nodes.

2.3 Implementation and Test

To demonstrate the functionality a small prototype series of network interfaces was produced and tested to meet the demands. An integration into two different propriatary hardware and software platforms were performed by two independent control equipment vendors. Applications tested using these platforms conluded the functionality in a heterogenious environment (Granum and Hansen, 1994).

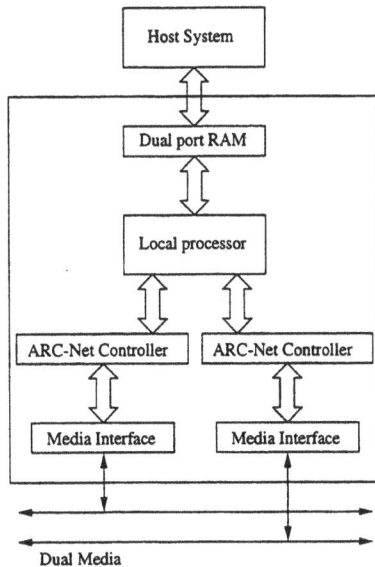

Fig. 1. Structure of network and network nodes.

3. REFERENCE MODEL

Used on an ISC system, a reference model paradigme, like [ISO TC184] (ISO-TC184 SC5 WG1, 1986), could be extended to cover items ranging from logistic to technical details. Our model is limited to cover the information flow in the distributed system. The functionality of the nodes/subsystems is described only to the extent needed to specify communication patterns and the traffic generated. A simple model is a key issue. To achieve simplicity the information which has to be supplied to describe a model element is limited to key parameters.

To ease the task of constructing a model of a target system, the reference model is structured as building blocks. The specification of a building block has to adhire to a set of rules which makes all blocks structural simular. Using this approach conformance classes for a generic set of standard components can be offered, leaving only modelling of more complex systems to the designer. The basic model elements offered to the designer are found to be:

- Network specifications.
- Specifications of connected equipment, including conformance classes for basic equipment and design guidelines for modelling complex equipment.

3.1 Network specifications

A target ISC system includes one or more network segments. For each segment information on replication strategi, selected bandwith and timing for reconfiguration has to be supplied. To make it possible to include other network technologies in the model also information on alternative topologies and medium access protocols can be supplied.

3.2 Specifications of connected equipment

In an ISC system conformance classes can be specified as generic descriptions of equipment as:

- GPS navigator.
- Autopilot.
- Heating systems.
- Power generators.

The conformance class for a specific type of equipment is expandable in a simple and structured way by using the rules given in the 'Design guidelines' for complex systems. A specific piece of equipment is allowed to cover more conformance classes (ie. a combined GPS navigator and autopilot). To make an instance of a component of a conformance class the designer has to add naming and values for variable parameters.

Complex and individual systems (as the main engine) cannot be covered by the relative simple properties of conformance classes. For these systems the reference model specifies guidelines for description and integration.

3.3 Model proporties

A description of connected equipment covers at least following items:

Functional description:

- Functionality of the specific equipment.
- Functions offered to other equipment (as server).
- Functions required from other equipment (as client).
- Coding and semantics of communicated information.

Traffic description:

- Type, amount and timing requirements of network traffic generated.

The functional description serves mainly as a source of information for calculating the traffic information needed for simulation. The structure of the information, which has to be supplied, is selected close to the physical structure of the ATO-MOS network, imploying network adapters, data channels and eventually individual tasks executed in the equipment. Figure 1 shows the relationship between network, adapters, channels and tasks.

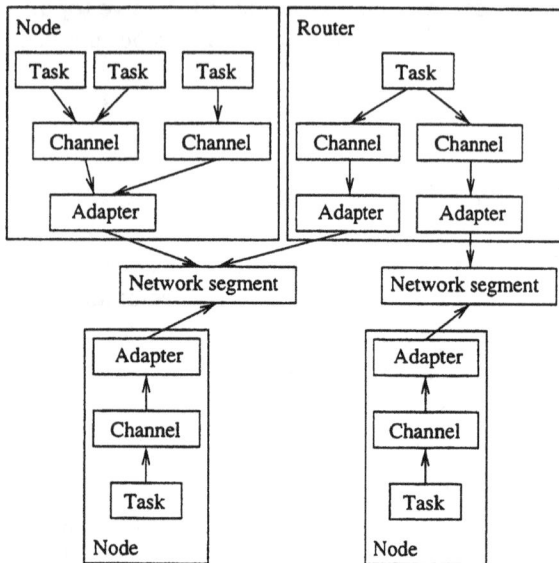

Fig. 2. Relationsship between modelling entities.

3.4 *Adapter*

Equipment is connected to the network through at least one network adapter. For each type of network adapter replication strategi, bandwith, timing parameters and buffer capacity has to be supplied. For each adapter used, information on naming, segment connection and addressing, has to be supplied.

3.5 *Channel*

All traffic is modelled as simplex data channels. A client server relationship is modelled as two channels, A data subscripsion service is modelled as a set of channels from the source to each of the destinations. This approach is selected due to the lack of a simple multicast service in the ATOMOS network, requiring that all multicasts are performed as series of point to point transmissions. For each channel imployed, source and destination together with characteristics and timing requirements of the dataflow has to be supplied.

3.6 *Task*

Normally, traffic is only specified on 'channel level'. In special cases, where the internal funtionality of af network node is wanted as a part of the model, modelling on 'task level' is possible. When modelling at task level traffic on the related channels is derived from the connected tasks. For each task imployed characteristics of scheduling, generated dataflow and timing requirements has to be supplied.

3.7 *Use of the Model.*

A model of a target system will include:

Structual information on network segments and equipment attached. Structual information on communication patterns in the system. Traffic information covering type, amount and timing requirements.

The structure and characterisics of the target system described by the model is used as input for the simulation process.

4. SIMULATION

The purpose of the simulation is to verify the behaviour of the communication system including the network segments and the attached adapters and data channels. For ISC systems especially the possibilty of simulating situations which cannot be runned "in vito" because of hazards etc. is vital.

For ISC systems normally three different working conditions are taken into account:

- Steady state at sea and no rolling.
- Manoeuvring in narrow water.
- Heavy weather.

These conditions will impose quite different activities in the ISC system. Simulations has to be done for each of these conditions including a variety of error conditions.

Two major assumptions for the simulation can be made based on the structure of the simulated system:

- As all protocol handling runs on a local processor all network characteristics are independent of host loads, making it possible not to include any host load parameters in the simulation of the network functionality.
- To make the simulation of data 'channels' independent of host load, data recieved at a node is always considered delivered instantaniously (ie. either consumed or discarded). Verification of the individual equipments ability to process recieved data is considered a separat task, normally covered by the vendor of the equipment.

These assumptions to a great extent decouples the simulation from the functionality of application software running in specific equipment, making the simulation task much more simple.

122

4.1 Simulator Input.

As input for the simulation, data extracted from the reference model and eventually recorded live data, are used.

The reference model sources following component information for a target system:

Network segment:

- Name and ID.
- Topology and MAC protocol used.
- Timing for reconfiguration in case of node adding/removal.
- Timing for regeneration in case of network errors (ie. token loss).
- Worst case access time, no network load (ie. token rotation time).
- Maximum token/network hold time.
- Maximum frame size.
- Bandwith (transmitted bit/second).
- Transmission overhead, total (ie. frame spacing, data frame headers and checksum, free buffer enquirement frame and acknowledgement frame)
- Replication strategi used.

Adapter:

- Name and ID.
- Network ID for attached network.
- Transmission setup time, average and max.
- Receiver latency.
- Bandwith for the controller (ie. bit/second or
- frames/sec).
- Buffer size on controller.
- Replicated network connection (A, B or A+B)

Channel:

- Name and ID.
- ID for source and destination adapter.
- Real-time requiremments (hard/soft/none).
- Periodic transmission: Period and max. allowed send skew.
- Non periodic transmission: Distribution (sporadic/burst) and average
- traffic (bits/second).
- Average packet length in each transmission.
- Send buffer space

Recorded live data can substitute traffic characteristics specified for selected equipment, refining the simulated runs and making comparison of simulator output and live data possible.

4.2 Simulator Output.

The simulator output is designed to focus on the modelled systems ability to meet the specified temporal behavoir. Output shall at least encourage:

- Statistics for individual or series of runs
- Traces for events where specifications were not met.
- Detailed trace of a simulation run.

The simulation is still on an experimental state. Preliminary work on two different approaches are in progress: An event based approach using modified timed colored Petri nets and a language orientated approach based on PROMELA/SPIN (Holzmann, 1991).

5. DISCUSSION

The ATOMOS network is based on exististing standards for MAC and application interface with redundancy and error handling added. The designed network interfaces, imploying LSI network controllers and local processor power, makes to a great extent network functionality host independent, providing better predictability and performance.

In ISC systems predictability and garantied proporties are key issues. In practice, for an asynchronious network, as the ATOMOS network, verification can only be done by simulation due to the complexity of the system. To serve as an infomation source for the simulation a reference model covering the information flow in the distributed system is developed. The level of details included in the model is a balance between the need of a fine granularity of the model and practical problems associated with specifying detailed information for a target system. It is found that the level of details included in our model is the outmost that can be expected for practical use.

Although final conclusions on the simulation cannot be drawn at this state, the prelimenary results shows that the concept is feasible as a tool for design as well as for verification purposes. Using simulations for verification implie that properties are presented on a statistical basis. It is found that this is sufficient for most purposes in an ISC system. In situations where formal guaranties for functionalities should be demanded, an approach using 'guaranties by design', as in time triggered systems, would be more feasible. (Koepetz and Grunsteidl, 1994)

6. REFERENCES

Dittmann, Kjeld (1992). Open automation architecture. *CAMS92.*

Granum et al (1993). *Detailed Description of Services and System Specification.* ATOMOS - Aalborg University.

Granum, M. and T. Hansen (1994). *ATOMOS Network Final Report*. ATOMOS-Aalborg University.

Hansen, T. N. and M. Granum-Jensen (1993). *Analysis of Basic Transmission Networks for Integrated Ship Control Systems*. ATOMOS - Aalborg University.

Holzmann, Gerard J. (1991). *Design and Validation of Computer Protocols*. Prentice Hall.

ISO-TC184 SC5 WG1 (1986). *The Ottowa Report on Reference Models for Manufactoring Standards*. ISO-TC184 SC5 WG1.

Jensen, T. and M. Granum (1993). *Analysis of Current and Proposed International Standards for Data Transmission*. ATOMOS - Aalborg University.

Koepetz, H. and G. Grunsteidl (1994). Ttp - a protocol for fault-tolerant real-time systems. *IEEE Computer p 14-23*.

Nielsen, J. Dalsgaard, K. Mølgaard Nielsen and N. Jørgensen (1995). Real-time communication networks onboard ships. *Cams95*.

CONTROL BLOCKS FOR DISTRIBUTED CONTROL SYSTEMS

Ronald Schoop and Heinz-Dieter Ferling

Schneider Automation, Steinheimer Strasse 117, 63500 Seligenstadt, Germany
E-mail: schoop@modicon.de, fg@modicon.de

Abstract: The industrial world lacks common models for the design and realization of distributed industrial-process measurement and control systems. The different models and approaches originate from private solutions of different vendors of control equipment aimed at different customers. An approach for distributed control blocks is introduced and refined, which describes distributed control systems, hierarchical control systems, and agent systems. Since this model is currently under standardization in the International Electrotechnical Commission, widespread use can be expected and further contributions from research institutes will be helpful.

Keywords: Control units, Design, Distributed control, Distributed models, Functional blocks, Industry automation, Model, Open control systems, Programming approaches, Programmable controllers, Real time communication, Standards

1 INTRODUCTION

In the industrial world distributed control systems are widely applied, but despite this huge application, common models for both the notation of distributed control functions and the design of the physical system are missing. The different models and approaches have come about by the different vendors of control equipment and by the different, initially targeted, customers: discrete manufacturing control (with the preference of using state diagrams) and process control (with the preference of using process variables).

In (Schoop and Strelzoff, 1995) the different programming models were compared. It was concluded, that an approach which uses distributed control blocks would bring advantages regarding a general execution model, a good granularity for mapping and a low communication overhead.
In this paper, the introduced approach will be refined and it will be shown, that this programming model is able to describe not only flat distributed systems, but also hierarchical systems and agent systems for both manufacturing and process control.

Since this model is currently under standardization in the International Electrotechnical Commission (IEC 1499-1, 1997), a widespread use can be expected.

2 GENERAL MODEL

The classical state machine model of Moore, Mealy, and Hufman was extended especially for programmable machines by splitting it into a **Control Unit** and an **Arithmetic Unit** (Burks, et al., 1963), see Fig. 1. Similar models, introducing an event mechanism for the Control Unit, have been published (Tisato and DePaoli, 1995).

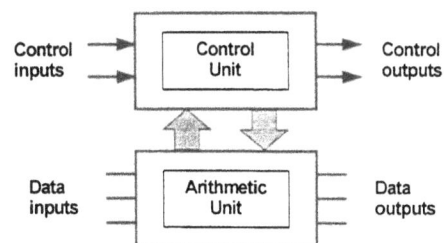

Fig. 1: Control Unit and Arithmetic Unit

This generic two layer model for programmable systems may be applied in two ways to distributed systems:

- The first approach takes this model as the overall model for the distributed system. In this case all internal synchronous execution is realized over networks, with a huge communication overhead required by the model.
- The second approach takes the two layer model as the description for any unbreakable node in the distributed system. This requires a new „third" layer for the notation of the whole system, but minimizes the communication burden, should an asynchronous model be used.

The control block model described below is based on the second approach.

3 IEC 1499 MODEL

3.1 Overview

Part 1 of the draft Function Block Standard (IEC 1499-1, 1997) introduces the architectural elements for the design and realization of industrial applications in a distributed system environment.

The model described in the Standard defines so-called "applications" as networks of interconnected function blocks. These applications are distributable onto sets of devices, which themselves are interconnected by communication networks, and which contain the necessary resources providing the execution environment for the function blocks in a distributed automation system. Besides the runtime support for executing the function blocks themselves, the execution environment also includes all interfaces to the communication networks and all interfaces between devices and the physical process.

Due to the fact that the term "function block" is so much overloaded in the context of industrial automation systems, the term "control block" is consequently used in the following text instead of IEC 1499 "function block" terminology. This avoids misinterpretations of the terminology of the Programming Languages Standard for Programmable Controllers (IEC 1131-3, 1993). The function block object and other language constructs introduced in that Standard are only used inside the sequential execution environment of IEC 1499 control block algorithms.

3.2 Control block types

Control blocks according to the IEC 1499 model are instances of control block types in an application.

The type specification of a control block defines all the characteristics which are common to all instances of that type (Fig. 2).

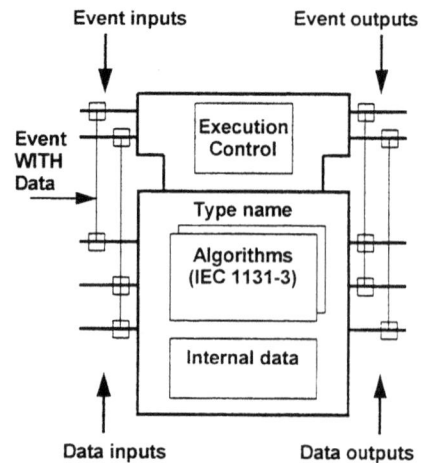

Fig. 2: Control block type

Control block types are characterized by two kinds of attributes. On the one hand there are attributes which define the interface part of a control block. which is visible from outside the block. These attributes are:

- the type name of the control block:
- the names and the graphical arrangement of event inputs and event outputs:
- the names, the data types and the graphical arrangement of data inputs and data outputs:
- the WITH relationships between event inputs and data inputs and event outputs and data outputs respectively.

On the other hand there are attributes which specify the internal function of a control block. including internally defined data, both of which are not visible from outside the block. The function is expressed by

- the execution control state machine and
- the algorithms of the block type.

Both are associated to events and data.

The execution control state machine handles incoming and outgoing events and. dependent on the actual internal state of the block, schedules algorithms. These algorithms operate on the values of input data, output data. and internal data to produce new values of output data and internal data.

3.3 Control block instances

Control block instances (or control blocks) are created as instantiations of control block types e.g. in an application. Such an instance is formed by a named, individual copy of the data structures specified by the control block type. The instance name is the key attribute of a control block.

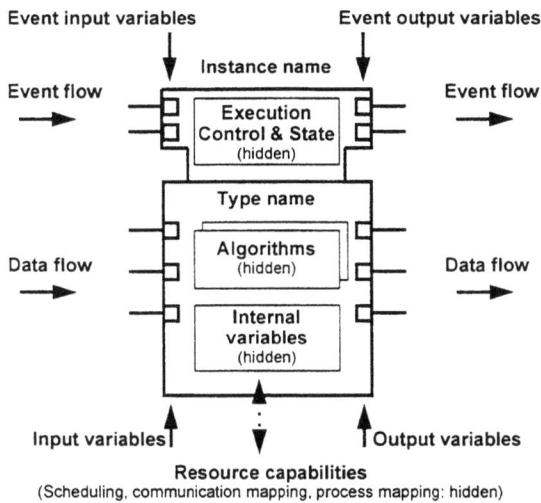

Fig. 3: Control block instance

As shown in Fig. 3, a control block instance comprises all variables which are necessary to allow persistent storage of instance data from one invocation of a control block to the next: event input and event output variables, execution control state, input variables and output variables, and internal variables carrying the internal state generated by the control block algorithms.

The execution of the algorithms is scheduled by the underlying resource which also provides access to the communication and process interfaces through specific service interface blocks (IEC 1499-1, 1997).

3.4 Control block execution model

The execution control state machine of a control block is started by an incoming event. Depending on the overall internal state of the block instance, the state machine decides which of the block type's algorithms is scheduled next for execution. At the same time all input variables which are associated with the incoming event in the type specification of the block are sampled to guarantee stable input data for the algorithm execution.

After an algorithm has been started by the resource, it performs its operations on the internal and output variables. Termination of the algorithm execution stimulates the issue, if any, of an output event associated with the algorithm. Associated output variables may be sampled, but normally input variable sampling is sufficient to guarantee data consistency for algorithm execution. Some specific issues are discussed in (IEC 1499-1, 1997).

The execution state machine decides if another algorithm has to be executed; otherwise the block execution is terminated. Note that algorithms of a control block are never executed concurrently. Therefore only the data sampling operations have to

be executed in critical regions. However, algorithms in different control blocks can be executed concurrently if correspondingly triggered by events.

3.5 Control block networks

Control block instances are interconnected to form control block networks (Fig. 4). A set of interconnected block instances may constitute an application or a composite control block type (see section 3.6).

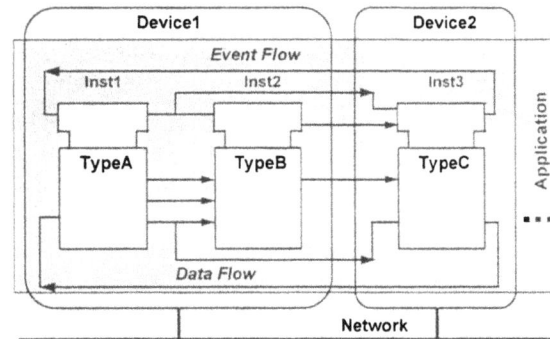

Fig. 4: Control block network

Blocks are connected by event connections which model the event flow in the control block network and through data connections which model the data flow in the control block network. Whereas event connections can be split (parallel triggering) and joined (multiple trigger sources), data connections can only be split (single data source model).

Fig. 4 also shows the mapping of the control blocks of an application onto different devices (for simplicity, one resource is assumed per device). Since the internal details of a control block are hidden from any application utilizing it, a control block forms the atomic unit of distribution over several resources or devices. This means that all elements contained in a given control block instance are configured to execute on the same resource. See the following section for an exception of this rule.

3.6 Basic and composite control block types

IEC 1499 introduces several kinds of control block types. The first kind, the **basic control block type**, was already discussed to some extent in section 3.2.

What needs to be mentioned additionally is, that the design of such a block type must specify the event and data interfaces, the execution control state machine (using a graphical or textual execution control chart language) and the algorithms of the block (using e.g. IEC 1131-3 programming languages).

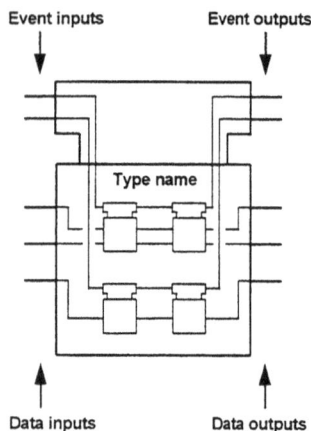

Fig. 5: Composite control block type

The second kind, the **composite control block type**, is shown in Fig. 5. This kind of block type allows the hierarchical decomposition of applications. Event inputs and outputs of the composite block interface are connected to event inputs and outputs of component block instances; data inputs and outputs are correspondingly connected. The "algorithm" part of the composite block may contain arbitrary control block networks.

Component blocks may again be of composite block type. Since composite blocks do not directly carry local state information, but only encapsulate it in component blocks, it is allowed to distribute these component blocks over several resources or devices.

3.7 Real time communication

An important aspect of the control block model is the mapping of the implicit event and data connections in already configured applications (and composite blocks) onto existing explicit real time communication facilities.

The Standard (IEC 1499-1, 1997) introduces a generic communication model which allows this mapping but many issues are left implementation-dependent or for definition in other Standards. To fulfill the mapping process, a design system for control block networks has to provide additional information for a connection, or a set of connections, which cross device borders, e.g. in a connection description table where every row contains corresponding attributes as shown in the following two samples:

Source	Destination	Data type	Trigger
...	...	...	...
Inst1.out	Inst2.in	INT	all 100ms
Oper.on	Motor.on	BOOL	Oper.cmd

The two examples above refer to the two commonly used communication patterns: periodic transfer of data with a given time period, or event driven transfer. Other mechanisms also exist, e.g. free running cyclic transfer or fetching data on demand.

Fig. 6 shows the realization of a connection for periodic data transfer. This mechanism can be used in an environment where control blocks which are configured to different devices are running cyclicly

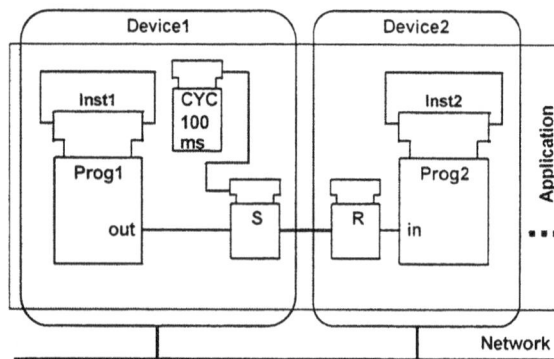

Fig. 6: Realization of periodic data transfer

and asynchronously. The cyclical behavior is achieved by connecting the "completion" event of each block back on to its start event (a necessary "initialization" event and its stimulation is not shown). Independent of the block cycles, the communication blocks S and R, which implement e.g. a unidirectional data transfer over the underlying communication channel, are driven by a 100ms trigger event source. Data consistency is guaranteed by input variable sampling as described in section 3.4, with the scan cycle of block Inst2.

Fig. 7 shows the realization of a connection for an event driven data transfer. The operator interaction (Oper) control block instance generates an event at its „cmd" event output, which directly triggers the communication block S to transfer data to its partner block R. The output event of this block, which is resource-initiated, triggers the Motor block instance, hereby starting the corresponding action.

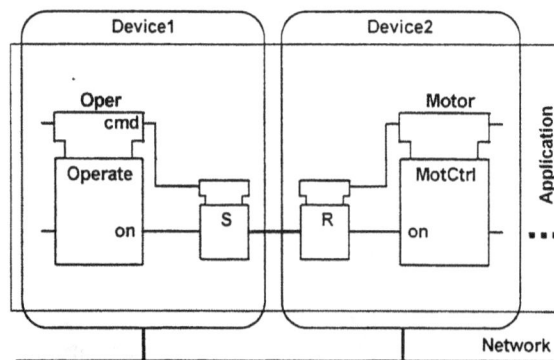

Fig. 7: Realization of event driven data transfer

The two examples not only show two different mechanisms to "realize" control block connections, but also demonstrate the versatility of the control block model. It is suited to model conventional cyclicly and periodically executing and interacting controller programs as well as systems which directly use "application" events generated by control blocks to trigger the activities of control blocks connected to them.

4 APPLICATIONS

As discussed in the chapter above, the initial intention for the IEC 1499 control blocks was the modeling of distributed control systems. Additionally, this model also allows the modeling of hierarchical systems and flexible agent systems. The following sections describe how this can be done.

4.1 Hierarchical distributed systems

The usual approach to model hierarchical distributed systems is based on nested function blocks where some of the function blocks are executed on remote devices and where "placeholders" of the remotely executing function blocks are located inside the Master function block. Fig. 8 presents an example.

How can the control block model be used to describe these kinds of systems ?
In the case of notation showing implicit communication, no "placeholder" will exist inside the Master control block, as shown in Fig. 9.

A composite control block would contain all control blocks of the Master block. The communication is hidden "inside" the event connections for „close" and „closed".

If we apply the notation showing explicit communication, composite control blocks may be

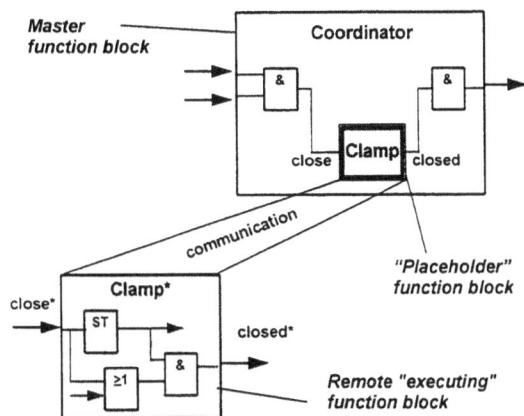

Fig. 8: Remote function block

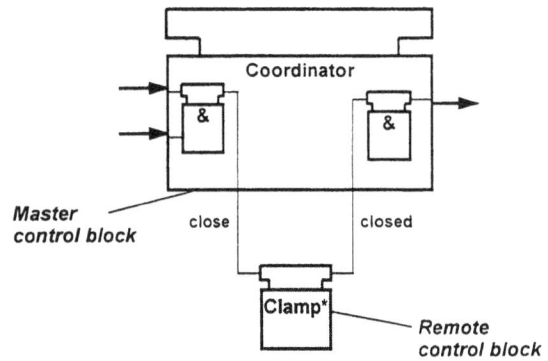

Fig. 9: Remote control block with implicit communication

used as "placeholders" for providing all communication capabilities, both inside the Master control block as well as inside the Remote control block, as shown in Fig. 10.

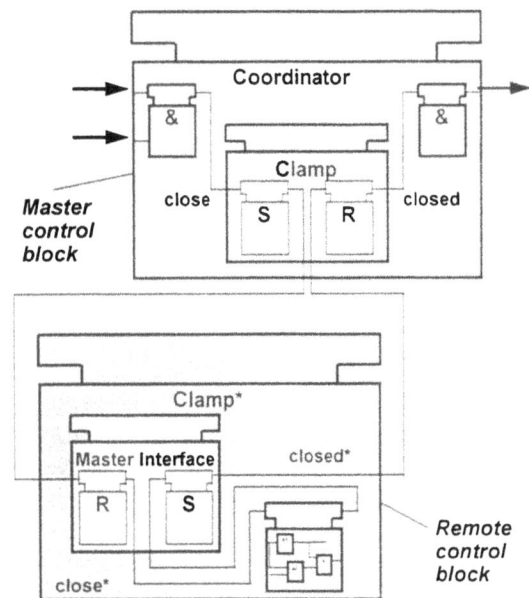

Fig. 10: Remote control block with explicit communication

These "placeholder" control blocks may be generated automatically and may be hidden to the user. However, for debugging purposes they may be shown.

4.2 Agent systems

Usual distributed control systems provide nodes whose control algorithms may change with time, but whose interconnections are pre-programmed at design time. To provide additional flexibility for the interconnections between the nodes, agent systems have been introduced in combination with conventional control systems, e.g. Programmable Controller systems. Typical applications for agent systems are intelligent manufacturing systems.

The main advantage of such agents is the **flexibility at run-time**:

- New agents may instantiated.
- Agents may fail and can be replaced.
- New inter-connections may be established.

Fig. 11 presents the modeling of agents with the control block model introduced above.

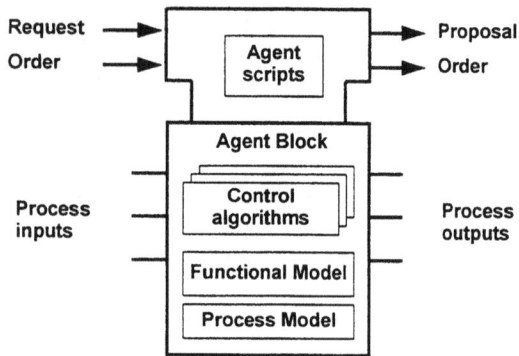

Fig. 11: Agent control block

The agent block accepts the events for orders (from other agents or conventional controllers) and requests (from other agents) and generates events for proposals (for other agents) and orders (to other agents or conventional controllers). The behavior of the agent block is defined by agent scripts which operate on a functional and a process model, and which activate the specific control algorithms. Such an agent block may be described as a composite block which can be refined into several elementary control blocks, one of them representing the conventional Programmable Controller functions.

A usual communication pattern between agents is the broadcast or black-board method. This is difficult to model with fixed event connections between agent blocks, at least using a graphical representation. Either textual representations like in section 3.7 (showing "Destination = Broadcast") have to be used for this kind of application, or the kinds of events have to be defined on both sides and the connections between the blocks are established dynamically at run time.

5 CONCLUSIONS

The control block model provides a unique methodology for handling distributed control systems. Its most important properties are:

- Already in the design phase the clusters of synchronized execution (inside the blocks by definition and between the blocks expressed by event connections) and of asynchronous execution (between the blocks in general), support an optimized realization.

- The model is not limited to distributed system design, but also entails the capability to model hierarchical and agent systems.

The presented model and results do not as yet describe a complete solution but show important steps in harmonizing the different worlds of industrial applications. A list of open issues for further research work still exists:

- Concurrency and data consistency in an asynchronous execution environment.
- Extension of known mapping/scheduling algorithms for blocks with event and data interdependencies.
- Extension of the model for redundant systems.
- Consideration of specific requirements of the world of process control systems.
- Design of tools for simulation of control defined by control blocks.
- Automatic generation of communication services for several different fieldbusses.

6 ACKNOWLEDGEMENTS

The authors would like to acknowledge the intense support in proof-reading and the resulting helpful contributions of Bryn Travers.

REFERENCES

Burks, A.W.; Goldstine, H.H.; von Neumann, J. (1963): Preliminary Discussion of the Logical Design of an Electronic Computing Instrument. In: Collected Works of John v. Neumann, Vol. 5. Macmillan, New York 1963.

IEC 1131-3 (1993): International Standard - Programmable Controllers - Part 3: Programming Languages. IEC, Geneva.

IEC 1499-1 (1997): Second Committee Draft - Function Blocks for Industrial-Process Measurement and Control Systems - Part 1: Architecture. IEC, Technical Committee 65, Working Group 6.

Schoop, R.; Strelzoff, A. (1995): Programming Approaches for Distributed Control Systems. In: DCCS '95. Preprints of the 13th IFAC Workshop. Toulouse-Blagnac (France), Sept. 1995.

Tisato, F.; DePaoli, F. (1995): On the Duality between Event-driven and Time-driven Models. In: DCCS '95. Preprints of the 13th IFAC Workshop. Toulouse-Blagnac (France), Sept. 1995.

DIVERSE DESIGN OF DCS WITH COMPACT CONTROLLER

Ik Soo Park, Eun Gee Kim

KEPRI, KEPCO

Abstract : An electronic failure of equipment may result from failure events within I&C systems. These failures are unavoidable and have to be taken into account by means of a redundant design. Also it is necessary to guarantee a higher quality of safety signals by raising the redundancy level so as to forestall spurious actuation. In the nuclear power plant, diversified control systems shall be provided for back-up in case of unavailability of the normal controls. Diversity (hardware/software), the use of at least two different means for performing the same function, shall be used to avoid Common Cause Failure(CCF) of both redundant system. This paper presents a practical diverse design of redundant DCS for the Power Plant using compact controllers.

Keywords: Distributed control, Process control, Redundent, Back-up.

1. INTRODUCTION

Replacing the old control systems of power plants with digital control systems is now a common project around the world. Numerous digital control systems have been installed in many power plants. This paper focuses on upgrade project for digital control system designed tripple redundancy using compact controller for Seoul #4 Power Plant built in 1960's. Before the control system upgrade was implemented, the control systems were pneumatic controllers and transmitters.

Reliability of these pneumatic systems was decreasing and operations & maintenance costs were increasing due to the single points of failure parts and the controllers required tuning for these systems. In 1992, it was scheduled to modify the oil-fired boiler for gas combustion and replace the entire these old pneumatic systems with distributed digital process control systems.

131

2. DEVELOPMENT AND STUDY OF DCS

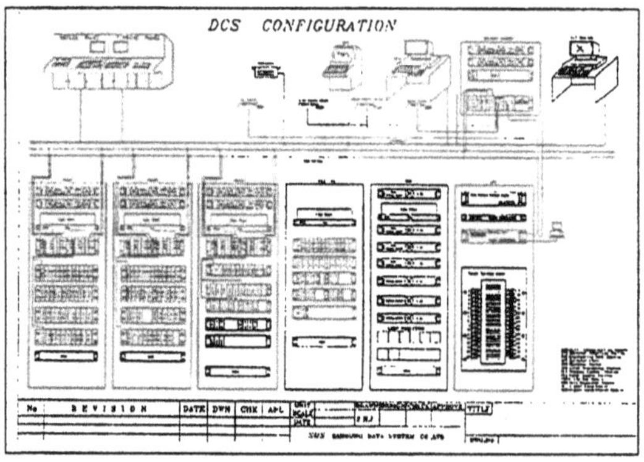

Fig 1. Seoul Power Plant DCS Configuration

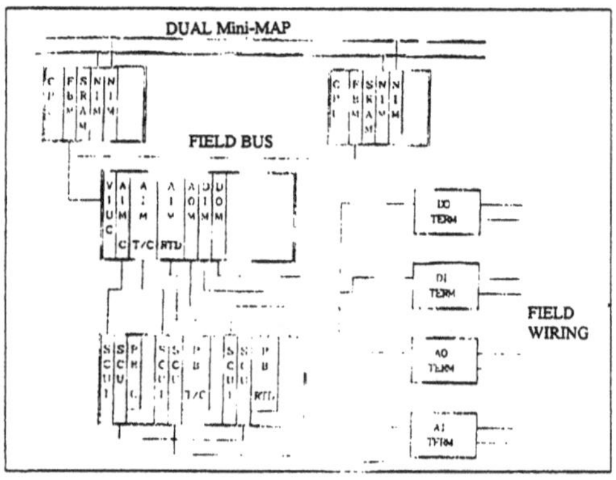

Fig 2. H/W Configuration in PCU

2.1 Survey of technology status
- Foreign technical status
- Boiler control systems
- Conventional digital control systems

2.2. System design
- System specification
- Design of system architecture

2.3. Development of PCS (Process Control Station)

Setup of system hardware
- Set up for dual power supply
- Design and setup of MFC (Multi-Function Controller)
- Design and setup termination, conditioning, conversion for signal
- Design and setup to interface H/A station

Development of software for PCS
- Implementation of on-line parameter change
- Implementation of downloading executable control file
- Implementation of field bus protocol
- Implementation of signal Input/Output function
- Development of function code
- Implementation of control configuration
- Implementation of I/O configuration

2. 4 Development of EWS (Engineering Work Station)

- Development of control configuration editor by Graphics
- Implementation of report configuration editor
- Implementation of trend configuration editor

2. 5 Development of OIS (Operator Interface station)

Setup of OIS hardware
- Design and installation of operator key board
- Design and setup to interface peripherals

Implementation of software
- Implementation of real-time data processing
- Implementation of historical data processing
- Implementation of alarm data processing

Implementation of process monitoring
- Alarm summary windows
- Panel overview window
- Trend display windows
- Group display window
- Parameter tuning windows

Implementation of process reporting
- Hourly, Daily, Monthly report
- Post-trip report
- Alarm report
- Group report

2.6 Development of system network

Development of Mini-MAP hardware and software
- Design and manufacturing of NIM (Network Interface Module)
- Implementation of Mini-MAP protocol
- Implementation of application program interface

Implementation of Ethernet interface
- Connection from Mini-MAP to Ethernet
- Implementation of workstation H/W and S/W

3. DIVERSE DESIGN OF DCS WITH COMPACT CONTROLLER

A distributed digital control equipment was used to imply the control system upgrade. The system consists of functional units (MFP, EWS, OIS) and Mini-MAP communication data highway. An expanded access for information systems is an Ethernet communication highway. A process control unit (PCU) is a part of the architecture to provide the power supplies, microprocessor, and I/O for the control system. Dual and automatic backup redundant, hot swappable power supplies reduce MTBF. For diversity design, the redundant functions and associated equipment including their support systems (e.g. power supply) should be independent from each other to limit the effects of a single failure on the redundant system. With respect to internal hazards, the independence of the redundant equipment

should be maintained by divisional separation of the redundancies. The separate I&C systems and equipment shall incorporate adequate diverse features to minimize the risk of Common Cause Failure. Therefore compact controllers including H/A station were installed at operator station in control room for divisional separation. The microprocessor of PCU performs continuous diagnostics and send 3 pulses per second to compact controller indicating normal operation. If compact controller should not receive 2 pulses per second from PCU, the contact for output in H/A station would be change from DCS to compact controller without bump. For bumpless of output, DCS watch the H/A station output and H/A station search the DCS output. Compact controllers for back-up control system have diverse hard-wired controls connected at the lowest I&C architecture level.

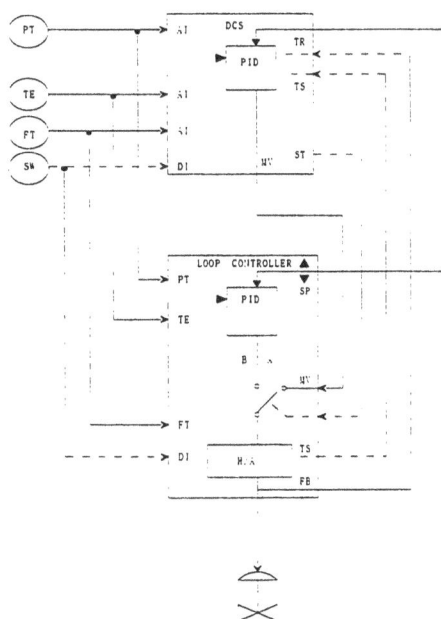

Fig 3. Connection of DCS and H/A station

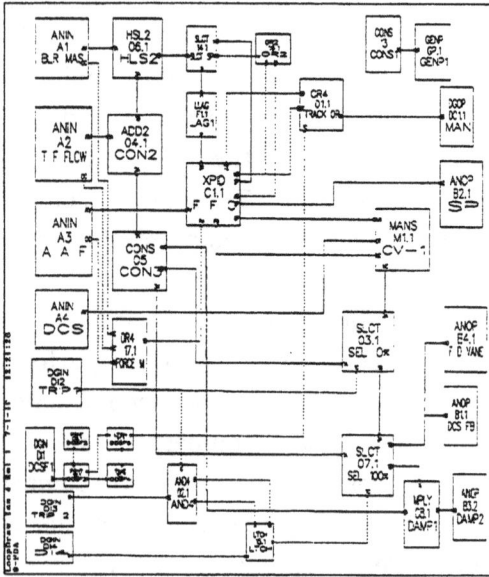

Fig 4. S/W for H/A station

4. RESULTS

Seoul #4 Power Plant upgraded the old pneumatic control system with tripple redundancy digital control system. Extreme care was taken during the design of control system to minimize digital control upgrade impacts.

Eventually, 42 compact controllers and four PCU cabinets were installed for diversity in Seoul Power Plant. In each PCU cabinets, there are redundant duel CPU boards called Master and Slave. During the start-up operation, we had experience of DCS failures twice, but the system maintained normal operation by Compact Controllers.

REFERENCES

Astrom, K.J. and Wittenmark, B. (1984), *Computer-Controlled Systems*, Prentice-Hall

Bailey (1991), *Bailey infi 90 manuals*, Bailey Controls Company

Bennett, S. (1988), *Real-Time Computer Control: An Introduction*, Prentice-Hall

Popovic, D. and Bhatkar, V. P. (1990), *Distributed Computer Control for Industrial Automation*, Marcel Dekker, Inc.

Rosemount (1989), *System 3 Manuals*, Rosemount Inc.

Siemens (1986), *Teleperm Me Manuals*, Siemens Aktiengesellschaft

Yamamoto, Shigehiko (1990), *New Advanced Control Technology in DCS - State of the Art and Their Future*, Yokogawa reference manual

Yokogawa (1989), *CENTUM-XL manuals*, Yokogawa Electric Corporation

ADOPTING MODERN COMPUTER SYSTEM TECHNOLOGY
TO NUCLEAR POWER PLANT OPERATIONS

Myung-Hyun Yoon * Ik Soo Park * W. Richard Pierce *
Rohini Mattu *** C. Dan Wilkinson ******

* KEPRI, 103-16 Munji-dong Yusung-ku, Taejon 305-380, Korea
** SAIC, 8682 Timberlake Road, Lynchburg, Virginia 24551, USA
*** SAIC, 11251 Roger Bacon Drive, Reston, Virginia 20190, USA
**** EPRI, 3412 Hillview Avenue, Palo Alto, California 94303, USA

Abstract: Instrumentation, control and monitoring systems in operating nuclear power plants generally utilize analog technology. As these systems age and become obsolete, utility companies are beginning to incorporate digital technology due to its proven record of high reliability in other industries, as well as wide spread availability. One of the major choices facing utilities in the I&C upgrade process is the use of digital computer-based systems to replace the existing analog control systems. Digital computer-based control systems provide improved functional control capabilities and the potential for communications with other plant control and monitoring systems. In addition, new challenges are associated with digital technology that must be faced by the utility. These include items such as evaluating the effects of electromagnetic and radio frequency interference on the system reliability, the need for software verification and validation, the review of abnormal conditions and events analysis (ACEs), as well as overall licensing aspects and the need to increase use of commercial-grade equipment to decrease cost, particularly for safety systems. This paper will review these aspects of adopting digital control technology and improvements in plant monitoring and network communications to nuclear power plants.

Keywords: Nuclear power stations, Control systems, Digital control, Communication networks, Man/machine interfaces

1. INTRODUCTION

Operating nuclear power plants were designed 20 to 40 years ago with analog instrumentation and control (I&C) technology. Today, most plants continue to operate with the original I&C equipment. This equipment is approaching or exceeding its life expectancy, resulting in increasing maintenance efforts to maintain system performance. Surveys of Licensee Event Reports in the United States show that a majority are related to I&C issues. Decreasing availability of replacement parts and the accelerating deterioration of the infrastructure of the manufacturers that support analog technology intensify obsolescence problems. As a result, operation and maintenance (O&M) costs are increasing.

Instrumentation and control systems in nuclear power plants need to be upgraded in a reliable and cost-effective manner to replace obsolete equipment, reduce operation and maintenance costs, improve plant performance, and enhance safety. The major drivers for the replacement of the safety, control, and information systems in nuclear power plants are the obsolescence of the existing hardware and the need for more cost-effective power production. Analog hardware that was designed 20 to 40 years ago is no longer fully supported by the original equipment manufacturer. Therefore the procurement of replacement modules and spares is costly, time consuming and, in some cases, not even possible. The increasing competition among power producers requires more cost-effective power production. The increasing operation and maintenance costs to maintain many of the analog I&C systems is counter to the needs for more cost-effective power production and improved competitiveness.

Technological improvements, particularly the availability of digital (computer-based) I&C systems, offer:

- Improved functionality, performance, and reliability;
- Solutions to obsolescence of analog equipment;
- Reduction in operation and maintenance costs; and
- Potential to enhance safety.

However when digital upgrades have been performed in nuclear power plants, problems with proprietary system architectures and new licensing and design issues have resulted in high implementation costs. There is a need for a systematic approach, leading to the identification, prioritization, and implementation of alternative I&C solutions in nuclear power plants. Viable alternatives range from extending the useful life of existing equipment to the complete and cost-effective system replacement.

Reliable, integrated information is a critical element for protecting the utility's capital investment and increasing availability and reliability. Integrated systems with integrated information can perform more effectively to increase productivity, enhance safety, and reduce O&M costs. A plant communications and computing architecture is the infrastructure needed to allow the implementation of I&C systems in an integrated manner. Current technology for distributed digital systems, plant process computers, and plant communications and computing networks support the integration of systems and information. However, even with the inherent technical advantages, digital systems will be implemented in nuclear power plants only if they support reduced power production costs and acceptance is achieved by the licensing authorities.

2. EPRI NUCLEAR POWER PLANT I&C UPGRADE PROGRAM

The Electric Power Research Institute (EPRI) and its member nuclear utilities are working together under the Integrated Instrumentation and Control Upgrade Program (EPRI, 1992d) on a three pronged approach to address I&C issues. The three prongs of the program consist of research and development activities, utility demonstration plant activities, and licensing stabilization activities. The research and development activities support the development and implementation of digital systems for cost and performance improvements, as well as providing a technical basis for qualification and licensing responses. It also provides part of the bases for the requirements and methodologies needed to design, develop, qualify, implement, operate, and maintain digital systems. The demonstra-

tion plant activities identify utility's needs, provide part of the bases for requirements and methodologies mentioned above, provide a test bed for and feedback on requirements and methodologies for upgrading systems, support the development of specifications for digital systems, and capture experience from implementing new digital systems. The licensing stabilization activities have provided technical support, as requested, for the industry licensing positions with the United States Nuclear Regulatory Commission (US-NRC) on digital systems which have been developed by utility working groups and facilitated by the Nuclear Energy Institute (NEI).

Part of the research and development activities, in conjunction with the demonstration plant activities, has been to define and develop a set of generic methodologies and guidelines that assist utilities in identifying, prioritizing, and implementing I&C solutions more effectively. The EPRI Instrumentation and Control Upgrade Program has developed a life-cycle management program for I&C systems. Life-cycle management involves the optimization of maintenance, monitoring and capital resources to sustain safety and performance throughout the plant life. Life-cycle management for I&C systems and components additionally may require the use of digital technology, when analog equipment cannot be cost-effectively maintained or when an improvement in performance is desired. The main product of the life-cycle management program is a set of methodologies and guidelines that, as part of the utility's overall life-cycle management effort, will enable nuclear power plants to fully consider I&C cost and performance improvements, including the application of digital technology. Specific examples of system specification and designs will also be developed through the application of the upgrade implementation methodologies to safety-related and non safety-related systems and system prototypes.

2.1 Planning Methodologies

Four strategic planning methodologies have been developed. The first two methodologies enable the utility to prepare an I&C life-cycle management program plan (EPRI, 1995d) and a plant communications and computing architecture plan (EPRI, 1994e). The last two methodologies enable the utility to perform long-term maintenance planning (EPRI, 1996b) and detailed upgrade evaluations (EPRI, 1996a) for I&C systems or components.

The Life-Cycle Management Plan is a long-term, strategic plan for managing the I&C systems over a selected planning period. The Life-Cycle Management Plan Methodology (EPRI, 1995d) guides a designated team of utility personnel through a comparison of I&C life-cycle management strategies and through existing and planned life-cycle management program activities to identify interfaces and integration options. On the basis of this comparison, the I&C Life-Cycle Management Plan is prepared. This plan includes:

- Identification of systems and components to be included in the program;
- Development of bases for upgrade or long term maintenance options;
- Initial cost and performance improvement estimates, prioritization for detailed upgrade evaluation, and deferred-upgrade maintenance planning; and
- Identification of related programs and organizational interfaces including key personnel and responsibilities.

The methodology is accompanied by a workbook which contains various outlines, worksheets, and generic interview questions and topics that aid in the development of a Life-Cycle Management Plan. The document describing the methodology also explains the overall process for planning and implementing the various elements of I&C life-cycle management, and the relationship of the other EPRI planning methodologies and guidelines. A plant-specific version of the life-cycle management plan is given in reference (EPRI, 1992b).

The Plant Communications and Computing Architecture Plan Methodology (EPRI, 1994e) provides utilities with a detailed set of instructions for preparing a Plant Communications and Computing Architecture Plan that will allow them to upgrade their I&C systems in a logical, cost-effective, and non-disruptive fashion. The Plant Communications and Computing Architecture Plan Methodology provides all of the information necessary to allow utilities to develop their strategic architecture plans in the most cost-effective manner possible. It guides a designated team of utility personnel through an assessment of the existing plant data network architecture, corporate communications architecture life-cycle management plans, and I&C life-cycle management implementation guidelines with respect to the communications architecture. On the basis of the assessment results, a Plant Communications and Computing Architecture Plan is prepared to address:

- Characterization of the existing network architecture;
- Characterization of the future network architecture in terms of a network model and communication standards for connectivity and interoperability of network elements;
- Set of network architecture requirements regarding the physical configuration, network access, network add-on provisions, network performance monitoring, and I&C equipment communications interfacing;
- Set of consistent human-machine interface guidelines for I&C systems; and
- Guidelines for process control equipment and computer platforms.

Some nuclear power utilities have used the need to upgrade their plant process computer (EPRI, 1992e) as an opportunity to develop a new plant communications and computing architecture. An example of a plant-specific architecture plan is given in reference (EPRI, 1993b).

The Systems Maintenance Plan Methodology (EPRI, 1996b) addresses long-term maintenance planning for systems or components where the initial screening in the Life-Cycle Management Plan indicates that detailed upgrade evaluation is not justified by cost and performance improvement potential, over the planning period. The Systems Maintenance Plan Methodology contains a process for developing a comprehensive System Maintenance Plan for each identified system. The Systems Maintenance Plan will present the most efficient approach for maintaining the operational goals and life expectancy of the system. The Systems Maintenance Plan Methodology will describe how to:

- Develop long range maintenance objectives,
- Baseline and analyze the existing maintenance process,
- Analyze failure rates, inventory practices, and obsolescence issues, and
- Implement maintenance related problem solving techniques.

The System Upgrade Evaluation Methodology addresses (EPRI, 1996a) a detailed evaluation of the I&C system when upgrading is indicated by the cost and performance screening in the Life-Cycle Management Plan. The Upgrade Evaluation Methodology is used to analyze each candidate system upgrade to determine if the upgrade is justified from a cost/benefit perspective. The Upgrade Evaluation Methodology is used to produce an Upgrade Evaluation Report for each candidate upgrade. The Upgrade Evaluation Report describes high level system functionality, upgrade alternatives and associated cost/benefit evaluations, and the recommended alternative. The upgrade evaluation process includes detailed cost and performance analysis; conceptual design options analysis; cost/benefit analysis; and upgrade recommendations. Analysis of conceptual design options includes the consideration of digital design basis changes, associated technical specification changes, and equipment selection candidates. If the system is to be upgraded, the Upgrade Evaluation Report forms the basis for the Functional Requirements Specification.

2.2 Integrated Plant Systems

As tasks become more complex, involving large numbers of subsystem interrelationships, the potential for human error increases. Therefore, reliable, integrated information is a critical element for protecting the utility's capital investment and increasing availability and reliability. Integrated systems with integrated information access can perform more effectively to increase productivity and enhance safety. Traditionally, systems upgrades have been implemented in a stand-alone manner, which has resulted in increased operation and maintenance costs. The modern technology available for distributed digital systems, plant process computers, and plant communications and computing networks is fully capable of supporting integration of systems and information. In fact, this capability has been proven in other process industries and in nuclear power plants outside of the United States.

Integration of the plant systems and information is essential to cost-effectively enhance cooperation between systems and to reduce unnecessary duplication of functions and information. The objectives of integrating plant systems and information are to:

- Improve plant availability and reliability,
- Reduce operations and maintenance costs,
- Reduce safety challenges, and
- Improve performance with existing and new equipment systems.

The plant communications and computing architecture of the plant supplies the infrastructure which allows the integration of systems and information. This infrastructure supports integrated upgrades, provides access to all of the plants information sources, and facilitates common interfaces between the human and the machine. This architecture will support the interoperability of systems and the interchangeability of equipment. It will also be designed to be easily expandable. This architecture is defined by a plan that includes a migration strategy to get from the current plant architecture to the final, desired architecture.

2.3 Issues Regarding Computer-Based Systems

Design and licensing issues have prevented utilities from benefiting from the cost and performance improvements possible with digital technology. Examples of the areas of concern for digital systems in nuclear power plants are licensing, software verification and validation (V&V), hardware qualification including electromagnetic interference compatibility and seismic, reliability, performance, separation, redundancy, fault-tolerance, common-mode failures, diversity, human-machine interfaces, and integration of systems and information through communications networks. Commercial-grade dedication of digital systems is an approach for more cost-effective implementations that is of considerable interest to the nuclear utilities. As part of the EPRI Instrumentation and Control Upgrade Program and other EPRI activities, approaches to address many of these concerns have been developed and the results are given in recent EPRI reports (1988, 1992a, c; 1993a, 1994a, b, c, d, f, g, h, i, j, 1995a, b, c, e).

The Guideline on Licensing Digital Upgrades (EPRI, 1993a) was developed to be consistent with the established process in the United States for plant changes under the US Code (10 CFR 50.59). It helps utilities design and implement digital upgrades, perform safety evaluations, and develop information to support licensing submittals. It suggests a failure analysis-based approach that encompasses digital-specific issues and other possible failure causes, addressing both according to their potential effects at the system level. Abnormal conditions and events (ACES) (EPRI, 1995a), as described in ANSI/IEEE ANS 7-4.3.2-1993 "Application Criteria for Programmable Digital Computer Systems in Safety Systems of Nuclear Power Generating Stations," play an integral role in this approach.

Guidance for electromagnetic interference susceptibility testing of digital equipment (EPRI, 1994b) and a handbook for electromagnetic compatibility of digital equipment (EPRI, 1994c)) have been developed. These reports integrate the current knowledge and understanding of the electromagnetic issues concerning the installation of digital equipment in power plants. They direct the utility toward practical and economical solutions for dealing with electromagnetic interference. The handbook also helps eliminate some misconceptions that questioned the reliability of digital equipment subjected to the electromagnetic environment of nuclear power plants.

Guidelines and a handbook for software V&V have been developed (EPRI, 1994d, 1995c, e). These products describe approaches to categorize the software systems in terms of importance and consequences of failure. They then identify levels of V&V consistent with these categorizations. The guidelines for V&V (EPRI, 1995c) developed a set of 16 V&V guideline packages based on the system category, development phase, and software system component which is being tested. For V&V methods in the guidelines that do not have a good description elsewhere in literature on how to use them, 11 sets of procedures have also been developed. The report identifies 153 V&V methods for software systems which can be used on the 52 identified software defect types. The guidelines developed were based on the attempt to identify the methods which were most successful in finding various types of defects, on the attempt to assure that the different guidelines catered to the different needs of different systems, and on the attempt to emphasize the practicality and cost-effectiveness of the methods recommended.

A process for the commercial-grade dedication of hardware has been developed (EPRI, 1988, 1994i) and proven very successful. The basic concepts of this process are being used as the starting point for proposed commercial-grade dedication processes for digital I&C systems. The use of commercial-grade programmable logic controllers (PLCs) for safety related systems is described in reference (EPRI, 1995b).

Guidelines for evaluating and dedicating commercial-grade PLCs have been developed (EPRI, 1994f, g). Considerable concern has been raised about annunciator systems and the magnitude of alarms that an operator must be aware of during a transient. The large number of alarms and the presentation of them make the operators job more difficult and can potentially contribute to human errors. Work has been done on more intelligent alarm systems and the methods for presenting them (EPRI, 1992a, c, 1994a). Additional areas that have been addressed are pressure transmitters (EPRI, 1994h), radiation monitoring systems (EPRI, 1994j), and wireless monitoring systems (EPRI, 1993c).

2.4 Demonstration Plant Projects

The utility demonstration plants essentially are the laboratories where I&C cost and improvement options are being researched and developed. There are five utility demonstration plant projects in progress which are providing the primary inputs, as well as testing, validation and refinement activities for the methodology and guideline development under the I&C Initiative.

Activities at each of the five demonstration plants may include the preparation of I&C life-cycle management plans and plant computing and control architecture plans; system screening, deferred-upgrade maintenance planning, and detailed upgrade evaluations; testing, validation, and refinement of various plant-specific methodologies and guidelines; and development of options and plans for integration of I&C cost and performance improvement activities with related life-cycle management efforts.

Demonstration project activities have taken place at the Tennessee Valley Authority's Browns Ferry Unit 2, Baltimore Gas and Electric Company's Calvert Cliffs Units 1 and 2, Northern States Power Company's Prairie Island Units 1 and 2, Entergy Company's Arkansas Nuclear One Units 1 and 2, and Omaha Public Power District's Fort Calhoun. An example of a plant-specific plan developed under the demonstration program is the architecture plan for the plant data network at Browns Ferry (EPRI, 1993b).

3. APPLICATION TO KORI-2 PLANT

Recognizing that the EPRI I&C Upgrade initiative could have potential significant benefits for the operation of nuclear power plants in Korea, the Korea Electric Power Research Institute (KEPRI) has entered into an agreement with EPRI to adopt the I&C planning and evaluation methodologies for use in evaluating the I&C systems of Korea Electric Power Corporation's (KEPCO) nuclear power plants. Science Applications International Corporation (SAIC) has supported EPRI in the development of the methodology

in the U.S., and will provide the initial support and technology transfer to KEPRI. A three phase effort has been adopted that will provide KEPCO with plans and reports on the I&C systems at Unit 2 of the Kori Nuclear Plant, and at the same time, provide for a transfer of the technology to KEPRI. Application of the methodology to other KEPCO nuclear plants may then be made by KEPRI.

The first phase of the effort was initiated in mid 1996, when EPRI and SAIC performed an initial scoping study of the I&C systems at Kori-2. Based on an initial review of the various systems, twenty were selected for further evaluation. The remainder of the efforts in phase 1 have now been contracted with EPRI and SAIC and are now underway.

Interviews with plant and utility personnel and reviews of plant documentation and records will result in status summaries of the twenty selected I&C systems. Maintainability assessments on the systems will provide the basis for a preliminary categorization into systems that are candidates for upgrade and systems that can be maintained for the planning period. This information is incorporated into the Life Cycle Management Plan. Responsibility for these activities are shared between SAIC and KEPRI in order that KEPRI can acquire the appropriate experience. EPRI provides a review function and the benefit of the experience from the demonstration plants. In addition to the system status summaries, subjects such as man-machine interfaces, plant networks, and computing platforms are documented. In order to prepare for potential replacement of plant monitoring systems, the Kori Plant Communications and Computing Architecture Plan will be produced as a part of the Phase 1 effort.

The second phase of the effort is a transition phase in which responsibility for the evaluations shifts from SAIC to KEPRI. Two upgrade candidate systems and two systems to be maintained are selected from the initial 20 systems. SAIC will provide the first system upgrade evaluation and Upgrade Evaluation Report. KEPRI will assume the lead role for the second system. Maintenance Plans will be developed for each system that is to be maintained using the same philosophy. KEPRI may then develop additional upgrade evaluations and maintenance plans with assistance from SAIC and EPRI.

In Phase 3, the Upgrade Evaluation Reports are used to develop functional specifications that are used in the procurement of replacement systems. The results of the process will eventually be realized when upgraded I&C systems are installed at Kori-2 that provide increased performance and lower O&M costs.

4. CONCLUSIONS

The implementation and integration of digital I&C systems enhances the ability to achieve the goals of improved availability and reliability, enhanced safety, reduced operations and maintenance costs, and improved productivity in nuclear power plants. The plant communications and computing architecture provides the infrastructure which allows the integration of systems and information. The modern technology of distributed digital systems, plant process computers (both monolithic and distributed), and plant communications and computing networks have proven their ability to achieve these goals in other industries and in nuclear power plants in other countries. The use of this modern, proven technology is a key contributor to improved competitiveness in nuclear power plants. EPRI has established an Integrated Instrumentation and Control Upgrade Initiative to support its member nuclear utilities in developing strategic plans and taking advantage of this modern technology to improve nuclear power plant competitiveness. KEPRI is beginning the process to apply this methodology to the Korea nuclear program.

5. REFERENCES

EPRI (1988). Guideline for the Utilization of Commercial Grade Items in Nuclear Safety Related Applications (NCIG-07). EPRI NP-5652.

EPRI (1992a). Alarm Processing and Diagnostic System. EPRI TR-100838.

EPRI (1992b). Browns Ferry Instrumentation and Control Upgrade Methodology. EPRI TR-101963.

EPRI (1992c). Control Room Alarm System Upgrades. EPRI TR-100584.

EPRI (1992d). Integrated Instrumentation and Control Upgrade Plan. EPRI NP-7343 Revision 3.

EPRI (1992e). Plant Process Computer Upgrade Guidelines, Vols. 1-3. EPRI TR-101566.

EPRI (1993a). Guideline on Licensing Digital Upgrades. EPRI TR-102348.

EPRI (1993b). Process Data Network Architecture Plan for the Browns Ferry Nuclear Plants. EPRI TR-103445.

EPRI (1993c). System Specification for the Wireless Programmable Process Monitoring System. EPRI TR-102287.

EPRI (1994a). Functional Specification Requirements for a Microprocessor-Based Annunciator System. EPRI TR-102872.

EPRI (1994b). Guidelines for Electromagnetic Interference Testing in Power Plants. EPRI TR-102323.

EPRI (1994c). Handbook for Electromagnetic Compatibility of Digital Equipment in Power Plants, Vols. 1&2. EPRI TR-102400.

EPRI (1994d). Handbook of Verification and Validation for Digital Systems, Vols. 1-3. EPRI TR-103291.

EPRI (1994e). Plant Communications and Computing Architecture Plan Methodology Vols. 1&2. EPRI TR-10104129.

EPRI (1994f). Programmable Logic Controller Qualification Guidelines for Nuclear Applications, Vols. 1&2. EPRI TR-103699.

EPRI (1994g). Programmable Logic Controller Requirements and Evaluation Guidelines for BWRs. EPRI TR-103734.

EPRI (1994h). Review of Technical Issues Related to the Failure of Rosemount Pressure Transmitters Due to Fill Oil Loss. EPRI TR-102908.

EPRI (1994i). Supplemental Guidance for the Application of EPRI NP-5652 on the Utilization of Commercial Grade Items. EPRI TR-102260.

EPRI (1994j). Utility Experience with Major Radiation Monitoring System Upgrades. EPRI TR-104081.

EPRI (1995a). Abnormal Conditions and Events Analysis for Instrumentation and Control Systems, Vols. 1&2. EPRI TR-104595.

EPRI (1995b). Experience with the Use of Programmable Logic Controllers in Nuclear Safety Applications. EPRI TR-104159.

EPRI (1995c). Guidelines for the Verification and Validation of Expert System Software and Conventional Software, Vols. 1-8. EPRI TR-103331.

EPRI (1995d). Life-Cycle Management Plan Methodology Vols. 1&2. TR-105555.

EPRI (1995e). Verification and Validation Guidelines for High Integrity Systems. EPRI TR-103916.

EPRI (1996a). System Upgrade Evaluation Methodology, Vols. 1&2. EPRI TR-104963.

EPRI (1996b). Systems Maintenance Plan Methodology, Vols. 1&2. EPRI TR-106029.

COMPONENT-BASED DESIGN OF LARGE
DISTRIBUTED REAL-TIME SYSTEMS

H. Kopetz

Technische Universität Wien
Austria

Abstract: Large distributed real-time systems can be built effectively by integrating a set of nearly autonomous components that communicate via stable control-free interfaces, called temporal firewalls. A temporal firewall provides an understandable abstraction of the subsystem behind the firewall, confines the impact of most changes to the encapsulated subsystem, and limits the potential of error propagation. This paper describes desirable properties of an ideal component, discusses the component interfaces and elaborates on the architecture support services that are needed to enable a component-based design of large distributed real-time systems.

Keywords: Real-time systems, distributed systems, composability, temporal firewall, reusability, component.

1. INTRODUCTION

In many engineering disciplines, large systems are built from prefabricated components with known and validated properties. Components are connected via stable, understandable, and standardized interfaces. The system engineer has knowledge about the global properties of the components–as they relate to the system functions– and of the detailed specification of the component interfaces. Knowledge about the internal design and implementation of the components is neither needed, nor available in many cases. A prerequisite for such a constructive approach to system building is that the validated properties of the components are not affected by the system integration. This composability requirement is an important constraint for the selection of an architecture for the component-based design of large distributed real-time systems.

A real-time computer system is always part of a larger system–this larger system is called a *real-time system*. A real-time system changes its state as a function of physical time, e.g., a chemical reaction continues to change its state even after its controlling computer system has stopped. It is reasonable to decompose a real-time system into a set of subsystems called *clusters* (Figure 1), e.g., the controlled object (the *controlled cluster*), the real-time computer system (the *computational cluster*) and the human operator (the *operator cluster*). We refer to the controlled object and the operator collectively as the *environment* of the real-time computer system.

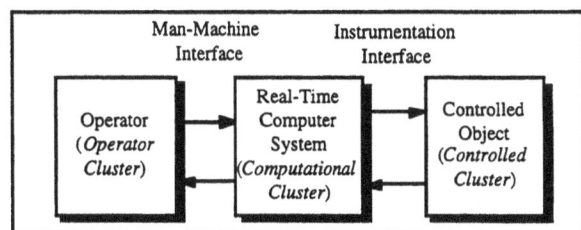

Figure 1: Real-time system.

If we freeze time, we can describe the current state of the controlled object by recording the values of its state variables at that moment. Possible state variables of a controlled object "car" are the position of the car, or the speed of the car. We are normally not interested in *all* state variables, but only in the *subset* of state variables that is *significant* for our purpose. A significant state variable is called a *real-time (RT) entity*. Every RT entity is in the *sphere of control (SOC)* of a subsystem (Davies 1978), i.e., it belongs to a subsystem that has the authority to change the value of this RT entity. Outside its sphere of control, the value of an RT entity can be observed, but cannot be modified.

The observation of RT entities in the controlled object is stored as a *real-time image* in the computer system. A real-time (RT) image is a *current* picture of an RT entity. An RT image is valid at a given point in time if it is an accurate representation of the corresponding RT entity, both in the value and the time domains. While an observation records a fact that remains valid forever (a statement about an RT entity that has been observed at a particular point in time), the validity of an RT image is *time-dependent* and thus invalidated by the progression of real-time.

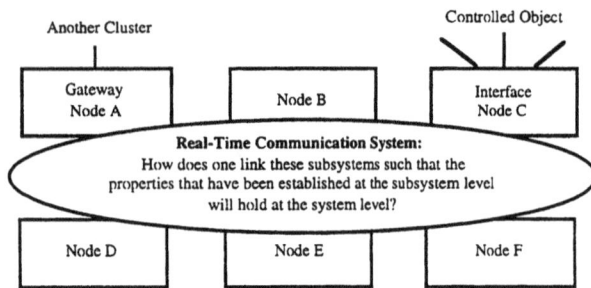

Figure 2: Structure of a computational cluster

We assume that the computational cluster of Figure 1 is structured as depicted in Figure 2. Computational nodes are connected by a distributed communication system to exchange messages. Some nodes, the interface nodes, have a second interface to the intelligent instrumentation. Gateway nodes connect one computational cluster to another computational cluster.

It is the objective of this paper to investigate what types of components, what properties of component interfaces, what type of communication system, and what support mechanisms from the architecture are required to apply the proven engineering principle of component-based design to the design of large distributed real-time systems.

The paper starts with a discussion of the required component properties. It then introduces a complete node—including hardware, software, and communication support—as the best choice for a component. The structure of such a reusable component is described and the component interfaces are investigated. Next, the concept of a temporal firewall is introduced to arrive at the desirable properties of an ideal component interface. In the final Section 5 the support services that must be provided by an architecture to enable the component-based design are discussed.

2. WHAT IS AN IDEAL COMPONENT?

In classic engineering disciplines a component is a self-contained subsystem that can be used as a building block in the design of a larger system. The component provides the specified service to its environment across the well-specified component interfaces. An example of such a component is an engine in an automobile, or the heating furnace in a home. The component can have a complex internal structure that is neither visible, nor of concern, to the user of the component.

Ideally, the larger system should be constructed from nearly autonomous components that can be integrated without violating the principle of composability: that properties that have been established at the component level will also hold at the system level(Kopetz 1997, p.34). Examples of such properties in the context of distributed real-time systems are timeliness and testability. An ideal component should be an autarchic unit that maintains its encapsulation when viewed from a number of different vantage points.

A unit of service provision: Most importantly, a component must be unit of service provision. The service is offered to the component environment across the service interface. In a distributed real-time system, the service consists of the timely processing and provision of the requested information. Since the validity of real-time information is time dependent (Kopetz and Kim 1990), the specification of the precise point in time when the information must be present at the component interface is of relevance. From the point of view of the service user, the internal structure of the component is irrelevant, as long as the specified service is provided at the anticipated points in time.

A unit for validation: It must be possible to validate the proper operation of a component in the value domain and in the temporal domain in isolation. The preconditions for the correct operation of the component, both in the domains of time and value, that must be satisfied by the component environment at the component/environment boundary must be precisely specified in the interface specifications. The specification of these preconditions is the input for the construction of an environment simulator that implements the testbed for the component validation. The post conditions that must be satisfied by a correctly operating component form the basis for the acceptance test of the component.

A unit of error containment: All errors that occur inside a component must be detected before the consequences of these errors propagate across the component interface. Otherwise, a defective component can falsify the operation of other components by the provision of corrupted output data across the component interface. Ideally, a component should support the fail-silence property (Schlichting and Schneider 1983): it either operates correctly (in the domains of time and value), is silent, or it produces a detectably incorrect result without disturbing the other components in the system. If a component has no other than such clean external failure modes, fault-tolerance can be achieved by replicating replica deterministic-components.

A unit for reuse: A component should be a unit for reuse. This requires that the component has standardized interfaces with some flexibility to support the integration of the component in diverse system contexts. For example, the component interface should support name translation to decouple the internal name space of the component from the external name space of the environment. This interface flexibility should not require a revalidation of the previously established component properties, such as correct operation or timeliness.

A unit of design and maintenance: Finally, a component should be a unit of design and maintenance. It is a well-known fact (Rechtin 1991) that system structures evolve along organization structures. If the work output of an organizational group is a nearly autonomous subsystem with well-specified interfaces, then the management of this group is simplified. The error containment boundaries around a component reduce the possibility of unforeseen consequences of software maintenance actions.

3. A NODE AS A COMPONENT

Considering the proposed properties of an ideal component, a complete node seems to be the best choice for a component in a distributed real-time system. A *node* is a self-contained computer with its own hardware (processor, memory, communication interface, interface to the controlled object) and software (application programs, operating system), which performs a set of well-defined functions within the distributed computer system.

In the analysis of an architecture for a component-based design identified four different types of components have been identified:

(i) Intelligent Transducers: An intelligent transducer is a component that consists of a sensor/actor element and an associated microcontroller that supports a standard interface to a fieldbus. An intelligent transducer hides the physical sensor or actuator behind the Transducer Interface (TDI), the fieldbus interface that always contains a temporally accurate measured value of the corresponding RT entity.

(ii) Interface Components: An interface component (see Figure 2 and Figure 3) is a component that has two interfaces, the Communication Network Interface (CNI) to the real-time network and the Controlled Object Interface (COI) to the controlled object. The COI supports the connection of one or more field busses to interface to the intelligent transducers.

(iii) Gateway Components: A gateway component is a component that has two CNI interfaces, each one to a different cluster that is connected by this gateway. The gateway component must select the information and transform the information representation of one cluster to that needed in the other cluster.

(iv) Computational Components: A computational component is a component that has a single CNI to the real-time network and performs some computational function in the system.

The interface component, the gateway component, and the computational components all have the hardware structure depicted in Figure 3. We call such a component a smallest replaceable unit (SRU).

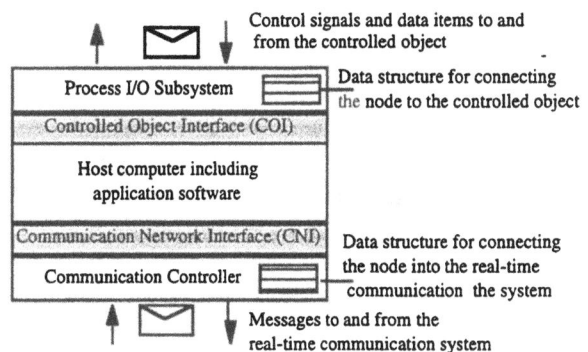

Figure 3: Structure of an SRU.

In general, the SRU hardware consists of a host computer, and one or two communication controllers (to the real-time bus and to the process I/O subsystem as shown in Figure 3). The host computer comprises the CPU, the memory and a real-time clock. The execution of the concurrently executing tasks within the host computer is controlled by the host operating system. The host computer shares the SRU internal communication network interface (CNI) with the communication controller and the controlled object interface (COI) with the process I/O subsystem. The host accepts the temporally valid input messages and produces the intended and timely output messages via the CNI and the COI.

The CNI is the most important interface in an architecture that supports a component-based design. Ideally the CNI is designed as a data-sharing interface free of control signals. The control signals for the communication system are generated autonomously by the communication controller according to a predesigned static communication schedule derived from the progression of time (a time-triggered (TT) schedule). Such an autonomous communication system helps to realize the composability of the architecture (Kopetz and Grünsteidl 1994). The CNI always contains the temporally accurate RT images of the RT entities in the controlled object. Such a data-sharing interface can be implemented in a dual-ported RAM, where each one of the interfacing subsystems can access the information from its side of the interface. The address of a RAM field corresponds to the static name and the contents of a RAM field contains the dynamic information of the interface data item at a particular moment of time. (At the chosen level of abstraction we disregard the low level control signals that are needed by the hardware arbiter of the dual-ported RAM. We assume that a data item is updated by an atomic operation).

If two systems communicate by exchanging messages with state-data semantics across a data-sharing interface, their coupling is looser than if they exchange event information, since the consumer does not have to consume every data item produced by the producer. In a state-based system, a consumer can read a produced value not at all, once, or many times. In state-based systems it is not required that consumer and producer proceed at the same rate.

The data structure within the communication controller (see Figure 3) describes the connection of the communication controller to the network and the required name translation between the component internal names and the external names. Furthermore it contains the dispatching information for the communication controller. In case the network is reconfigured or the component is used in a new context, only this data structure has to be modified. The software in the host computer is not changed when the SRU is reused in a new application.

4. TEMPORAL FIREWALLS

On the conceptual level, the CNI between the host computer and the communication network (or the controlled object) can be seen as erecting two unidirectional temporal firewalls (Kopetz and Nosssal 1997) that connect the component to its environment.

A temporal firewall is a unidirectional data-sharing interface with state-data semantics where at least one of the interfacing subsystems accesses the temporal firewall according to an a priori known schedule and where at all points in time the information contained in the temporal firewall is temporally accurate for at least d_{acc} time units into the future.

The subsystem that accesses the temporal firewall according to the *a priori* known schedule is called the

time-triggered (TT) subsystem. No control signal is crossing the temporal firewall. The information provider has to update the RT image in the temporal firewall according to the dynamics of the corresponding RT entity. If the information-providing subsystem ceases to operate, the information in the temporal firewall is invalidated by the passage of time.

4.1 Stable Properties of Temporal Firewalls

The following stable properties characterize a temporal firewall. Knowledge about these properties is available *a priori* to all interfacing subsystems:

(i) The addresses (names) and the syntactic structure of the data items in the temporal firewall. The meaning of the data items is associated with these names.

(ii) The points on the global time base when the data items in the temporal firewall are accessed by the TT subsystem. This information enables the avoidance of race conditions between the producer and the consumer. A race condition could lead to a loss of replica determinism in replicated temporal firewalls.

(iii) The temporal accuracy d_{acc} of the data items in the temporal firewall (Kopetz 1997, p.102). This knowledge is important to guide the information consumer about the minimum rate of sampling the temporal firewall. The absolute timepoints when the TT subsystem accesses the temporal firewall are reference points for the temporal accuracy of the information in the temporal firewall.

4.2 Obligations of the Subsystems

To ensure the proper information flow across a temporal firewall, the producer and the consumer subsystem must comply with the followings obligations.

Producer: The producer of the RT-images stored in the temporal firewall is responsible that the *a priori* guaranteed temporal accuracy of the RT-images is *always* maintained. It must update the state information with such a frequency that the guaranteed temporal accuracy is sustained even immediately before the point of update. In case the producer of the information is the TT subsystem, the producer is allowed to access the temporal firewall only at the *a priori* established time points t to avoid race conditions for access to the temporal firewall. In case the producer of the information is not the TT subsystem, the producer is allowed to access the temporal firewall at any point in time outside a *critical interval* around t. The duration of this critical interval is $[t-2g, t+2g]$, where t is the access time of the TT subsystem and g is the granularity of the global time(Kopetz 1997, p.55).

Consumer: Based on the *a priori* knowledge about the temporal accuracy of the RT images in the temporal firewall, the consumer must sample the information in the temporal firewall with a sampling rate that ensures that the accessed information is temporally accurate at its *time of use* of this information. The consumer is only allowed to access the information in the temporal firewall when it knows (based on the *a priori* knowledge) that the producer is not accessing it (see above). If the consumer violates these access constraints, replica determinism may be lost, or, in the worst case, the consumed information may be corrupted. (The implementation of protected shared objects can avoid information corruption, but cannot guarantee replica determinism).

4.3 Temporal Firewalls in the Validation Process

A temporal firewall is a small and stable interface that provides understandable abstractions of the relevant properties of the interfacing subsystems. Conceptually, the RT images in the temporal firewall are closely related to the image presented by a sensor of an analog RT entity in the environment. Temporal firewalls are thus based on an accustomed view of the world. A temporal firewall is a rigid interface with *a priori* known stable attributes. This rigidity is the strength and the weakness of the temporal firewall concept at the same time. It is a strength, because a precisely defined stable interface induces structure into an architecture. Since this structure is time-invariant it can be relied on when building the system and when reasoning about the properties of the system. System validation and error confinement are facilitated if subsystems are encapsulated within a rigid structure. On the other hand, rigid internal interfaces limit the flexibility and the adaptation of the architecture to a changing request pattern from the environment. In some cases, these stable internal interfaces can be the cause for a waste of system resources. Considering the two trends that determine the current evolution of real-time systems—improving cost performance and increasing complexity—we feel that a tradeoff in favor of reduced complexity is justified in many applications.

Preconditions and Postconditions. Assume a component that is encapsulated between two temporal firewalls. These two firewalls form the only interfaces of this component to its environment. The first firewall, the input firewall, delivers RT images from the environment of this component into the component under consideration, and the second one, the output firewall, delivers the results from this component to the rest of the system. The stable properties of the input firewall form important *preconditions* for the validation of the component under consideration. Many assumptions about the environment are contained in the specification of this input firewall. Since the information flow across a temporal firewall is unidirectional, there is no dependence of the producer subsystem at the producing side of the input firewall on the proper operation of the component under consideration. The stable properties of the output firewall form important postconditions of the validation. In the validation process it must be demonstrated that the *postconditions,* given in output firewall specification, are always TRUE, provided the preconditions associated with the input firewall hold.

Temporal firewalls partition a large distributed real-time system into a set of nearly autonomous subsystems with fully specified interfaces in the temporal domain and in the value domain. Each one of these subsystems can be developed and tested independently from the other subsystems. This systematic decomposition during the design phase and the ensuing constructive composition during the validation and integration phase facilitates the component-based development of large distributed real-time systems. The implementation of a synchronous time-triggered communication system, a means to implements the temporal firewalls, facilitates the formal reasoning about the relevant properties of distributed real-time architectures (Rushby 1997).

Error Containment Interface. A temporal firewall is free of control signals. Therefore there is no possibility of a control-error propagation across a temporal firewall. Since the information flow across a temporal firewall is unidirectional, a data error can only propagate from the producer to the consumer. There cannot be any error propagation from the consumer back to the producer. Thus a temporal firewall acts as an effective error-containment interface. It encapsulates a component and restricts the visibility of its internal mechanisms. The static temporal properties of the temporal firewall ensure that the temporal obligations of the partners of a client-server interaction are enforced (Kopetz 1996). Changes made inside a component do not effect the static properties of the temporal firewall. These changes are encapsulated within the component and cannot ripple through the total architecture.

4.4 An Example of a Temporal Firewall

Consider an air traffic control system consisting of four components, a data-acquisition component, a man-machine interface component, a collision avoidance component, and a route planning component (Figure 4).

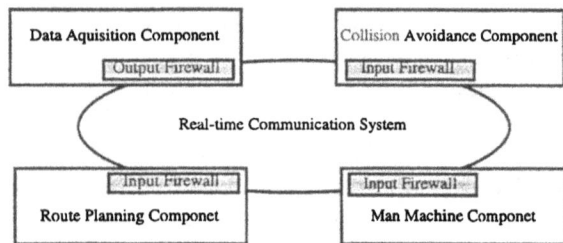

Figure 4: Components in the air traffic control system example.

The data acquisition component controls the radar stations and locates the object in the airspace. The results of this component are conveyed to the other components by a temporal firewall (Figure 5). This temporal firewall contains an RT image for every RT entity in the sky. These images are updated such that the temporal accuracy of the information is in conformance with the application requirement. The real-time communication system of Figure 4 replicates the temporal firewall of Figure 5 in each one of the other three components of the distributed system.

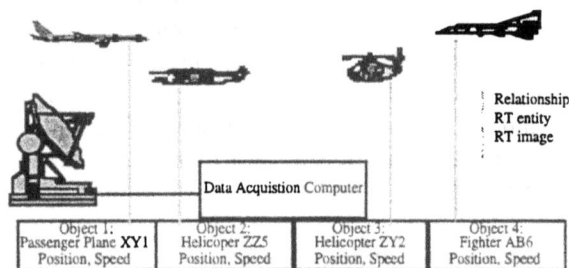

Figure 5: Temporal firewall of the data acquisition component.

5. ARCHITECTURE SUPPORT

Component-based design of large real-time distributed systems requires architecture support for the realization of the stated objectives. We have identified three areas that must be supported at the architecture level: clustering and composability, clock synchronization, and a responsive membership service.

5.1 Clustering and Composability

Large systems can only be built if the effort required to understand the system operation, i.e., the complexity of the system, remains under control as the system grows. The *complexity* of a system relates to the number of parts, and the number and types of interactions among the parts, that must be considered to understand a particular function of the system. The effort required to understand *any*

particular function should remain constant, and *independent* of the system size. Of course, a large system provides *many more different* functions than does a small system. Therefore the effort needed to understand *all functions* of a large system grows with the system size. The complexity of a large system can be reduced if the inner behavior of the components can be encapsulated behind stable and simple interfaces. Only those aspects of the behavior of a component that are relevant to the function under consideration must be examined to understand the particular function. This requires the architectural support for clustering and composability with rigid temporal firewalls between cluster.

5.2 Establishment of a Global Time Base

If the real-time clocks of all nodes of the distributed system were perfectly synchronized and all events were timestamped with this synchronized time, then it would be easy to measure the interval between any two events or to reconstruct the temporal order of events, even if variable communication delays generated differing delivery orders. In a loosely coupled distributed system where every node has its own local oscillator, such a tight synchronization of clocks is not possible. A weaker notion of a universal time reference, the concept of *global time*, is therefore introduced into a distributed system. Suppose that each node has its own local physical clock c^k that ticks with microtick granularity g^k. Assume that all of the clocks are internally synchronized (Kopetz 1997, p.52) with a precision Π, i.e., for any two clocks j,k and all microticks

$$\left| z(microtick_i^j) - z(microtick_i^k) \right| < \Pi.$$

$z(e)$ is the timestamp of event e generated by the reference clock z of an omniscient external observer. It is then possible to select a *subset of the microticks* of each local clock k for the generation of the local implementation of a global notion of time. We call such a selected local microtick i a *macrotick* (or a *tick*) of the global time.

The global time t is called *reasonable*, if all local implementations of the global time satisfy the condition

$$g > \Pi$$

the *reasonableness condition* for the global granularity g. This reasonableness condition ensures that the synchronization error is *bounded* to less than one *macrogranule*, i.e., the duration between two global ticks. If this reasonableness condition is satisfied, then for a single event e, that is observed by any two different clocks of the ensemble,

$$| t^j(e) - t^k(e) | \le 1,$$

i.e., the global timestamps for a single event can differ by at most one tick. *This is the best we can achieve.* Because of the impossibility of synchronizing the clocks perfectly, and the denseness property of real time, there is always the possibility of the following sequence of events: clock j ticks, event e occurs, clock k ticks.

5.3 Membership Service

The proper operation of many system level functions depend on the correct and timely operation of a minimal set of components. In case a component fails (silently) the other components must be informed within a specified time interval–the error detection latency–about this component's failure. This is the task of the *membership service*. A point in real-time when the membership of a node can be established, is called a *membership point* of the node. A small temporal delay between the membership point of a node and the instant when all other nodes of the ensemble are informed in a consistent manner about the membership, is critical for the correct operation of many safety-relevant applications.

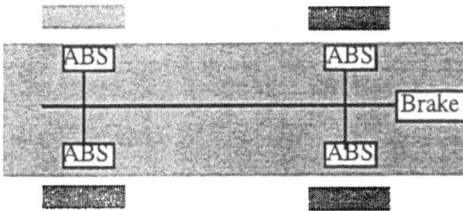

Figure 6: Example of an intelligent ABS in a car.

Consider an intelligent ABS (Antiblock System) braking system in a car (Figure 6), where a node of a distributed computer system is placed at each wheel. A distributed algorithm in each of the four computers, one at each wheel, calculates the brake-force distribution to the wheels, depending on the position of the brake pedal actuated by the driver. If a wheel computer fails or the communication to a wheel computer is lost, the brake-force actuator at this wheel autonomously transits to a defined state, e.g., in which the wheel is free-running. If the other nodes learn about the computer failure at this wheel within a short latency, e.g., a single control loop cycle of about 5 msec, then the brake force can be redistributed to the three functioning wheels, and the car can still be controlled. If, however, the loss of the node is not recognized with such a low latency, then, the brake force distribution to the wheels, based on the assumptions that all four wheel computers are operational, is wrong and the car will go out of control.

6. CONCLUSION

Component-based design of large distributed real-time systems requires computational components with stable and understandable interfaces, and an architecture that provides the necessary support services, such as clock synchronization and membership. The time-triggered architecture developed at the Technische Universität Wien is one example of a distributed real-time system architecture that provides the needed support services. In many embedded application, where the limited flexibility of such an architecture is of no major concern, the technical and commercial benefits gained from a systematic component-based design can be significant.

ACKNOWLEDGMENTS

This work was supported in part by the ESPRIT LTR project DEVA.

REFERENCES

Davies, C. (1978). Data Processing Spheres of Control. *IBM Systems Journal*. Vol. 17. pp. 179-198.

Kopetz, H. (1996). A Node as a Real-Time Object. *Proc. of the IEEE Workshop on Object Oriented Real-Time Systems*, Laguna Beach, Cal. IEEE Press. pp. 1-8.

Kopetz, H. (1997). *Real-Time Systems, Design Principles for Distributed Embedded Applications; ISBN: 0-7923-9894-7.* Boston. Kluwer Academic Publishers.

Kopetz, H. and G. Grünsteidl (1994). TTP-A Protocol for Fault-Tolerant Real-Time Systems. *IEEE Computer*. Vol. 24. pp. 22-66.

Kopetz, H. and K. Kim (1990). Real-Time Temporal Uncertainties in Interactions among Real-Time Objects. *Proc. 9th IEEE Symp. on Reliable Distributed Systems*, Huntsville, AL. pp. 165-174.

Kopetz, H. and R. Nosssal (1997). Temporal Firewalls in Large Distributed Real-Time Systems. Technische Universität Wien, Institut für Technische Informatik.

Rechtin, E. (1991). *Systems Architecting, Creating and Building Complex Systems*. Englewood Cliffs. Prentice Hall.

Rushby, J. (1997). Systematic Formal Verification for Fault-Tolerant Time-Triggered Architectures. *Proc. DCCA 6*, Garmisch, Germany. IEEE Press. pp. (Preprints) 191-210.

Schlichting, R. D. and F. B. Schneider (1983). Fail-Stop Processors: An approach to designing fault-tolerant computing systems. *ACM Trans. on Computing Systems*. Vol. 1. pp. 222-238.

EFFECT OF DYNAMIC STRUCTURING ON DISTRIBUTED CONTROL SYSTEM PERFORMANCE

A. Stothert* and I.M. MacLeod[*,1]

Department of Electrical Engineering, University of the Witwatersrand, WITS 2050, South Africa

Abstract: The performance of a distributed computer control system that uses communication both for structuring and for knowledge and information transfer between agents is discussed. This agent-centred approach relies on the use of agent templates and interaction rules — resulting in a distributed computer control system framework that is able to adapt as operating requirements change. Implementation of a distributed controller, which simulates a simple conventional controller is described. The effect on performance of this approach, both in terms of processor loading and network traffic is shown to be minimal.

Keywords: distributed computer control systems, agents, dynamic structuring, fault tolerance

1. INTRODUCTION

Distributed artificial intelligence is a rich research area (Decker 1987, Durfee 1991, Bird 1993) which concentrates on cooperation between autonomous entities called agents. An *agent* is an autonomous program that communicates with like agents to perform tasks on their behalf or in cooperation with them. Researchers have noted the important role of communication in distributed systems, but only as a support tool for the agents in the distributed system. However we suggest (Stothert and MacLeod 1995) that communication can also be used to structure the distributed system, resulting in a system that organises itself to suit the plant and to optimally use available resources.

This paper describes the implementation of a distributed system which uses communication for structuring purposes as well as knowledge/ information transfer between agents. The objective is to demonstrate the technique and to quantify

its effect on distributed system performance. The paper briefly introduces the theoretical framework and an implementation (section 2) before discussing a distributed control example (section 3). Section 4 discusses operation of the implemented controller and provides an indication of the overheads (communication and processor loading) associated with the developed framework.

2. BACKGROUND

Previous work (MacLeod and Stothert 1994) showed that the design of distributed computer controllers focusses on the distinction between a priori and operational knowledge. Briefly, a priori knowledge is that knowledge which is used to determine the control approach, the types of agents required and how they might interact. Operational knowledge is the knowledge that becomes available during plant operation and is used to determine future control actions.

We are mainly concerned with operational knowledge since it effects the communication system in a distributed system. In well-defined systems

[1] The support of the South African Foundation for Research Development and the University of the Witwatersrand is gratefully acknowledged.

operational knowledge only requires that data be transferred between fixed agents (Cohen 1991, Lesser and Corkill 1981). In more flexible systems operational knowledge is required to dynamically determine which agents should communicate, contract net and negotiation protocols are typical (Davis and Smith 1983, Sycara 1989). However, operational knowledge also includes information about which agents are required at a particular time. In other words, operational knowledge can be used to *add or remove* agents from the distributed system dynamically. The framework discussed below describes the design and implementation of distributed computer control systems that use operational knowledge to add and remove agents from the distributed system.

Work by Lueth (Lueth *et al.* 1995) demonstrates the performance advantages of dynamic agent/task allocation. The approach centres on a manual user task selection rather than the approach adopted here which focusses on the agents themselves using operating conditions to launch other agents. The distinction is important as the agent-centred approach results in characteristics like adaptability, fault tolerance and partial solutions which are not as evident in a centralised approach like the one advocated by Lueth.

In order to develop a distributed system which supports agents that can dynamically add or remove other agents it is necessary to introduce the idea of an agent template (Stothert and MacLeod 1995). The template is a "blue-print" from which actual agents can be implemented. Since the focus of distributed intelligent systems is on how agents interact or cooperate to achieve system wide goals it is necessary for agent templates to include not only functionality, data structure and input/output descriptions but also interaction rules.

The interaction rules describe the intended role and required behaviour of the agent. Werner (Werner 1989) shows that role can be used to describe cooperation between agents and that role can be expressed by *if... then...* rules. Similarly behaviour can be expressed in a rule format. In other words, each agent template includes a knowledge base which is used to develop the distributed system structure by examining the intended role and behaviour of an individual agent. For example a pump agent's role is to monitor and control a pump. The agent recognises that a possible fault condition has arisen. Using interaction rules that describe the pump agents behaviour another agent who's role is diagnosing faults is launched by the pump agent.

A prototype implementation of these ideas under Windows NT3.51 on a two node network (two

486DX2 PC's) has been developed. The implementation is aimed at examining the use of interaction rules as a design technique for distributed controllers, the dynamic launching and removing of agents and the performance issues associated with dynamic structuring.

3. A DISTRIBUTED PID CONTROLLER

In order to demonstrate the distributed system design philosophy outlined in section 2 and the operation of a dynamic distributed system, an adaptive PID controller (Franklin *et al.* 1986) was implemented. A PID controller was chosen as it is well known in the control community and can be adapted to suit the system being controlled. Note we are not proposing this implementation as an alternative PID solution.

The idea is to have a simulation agent which can produce different classes of feedback error signal and to implement a distributed PID controller which identifies the error signal class and assigns a corresponding controller structure, i.e., proportional only, integral only or proportional integral. Depending on the controller structure required different agents will be present in the distributed controller.

Recognising the requirements of a distributed PID controller, four agent templates can be defined. An agent to identify the required PID structure, an integrating agent, a differentiating agent and an agent to produce the PID control signal. The role of each of the agents is evident from the tasks they are required to carry out and is a result of a priori knowledge. Depending on operational knowledge, which will be used to identify the required PID structure, different agents will be present at different times. In addition to the agents for the PID controller an agent to simulate a plant error signal will also be present in the distributed system.

With the required agent templates identified, and the roles of each agent are understood, the behaviour of individual agents can be established by looking at the behaviour of the overall system. In this way a complete distributed system design is possible. The design for the PID controller is discussed in more detail in (Stothert and MacLeod 1997).

4. RESULTS

Several simulations were carried out using the PID distributed controller in order to illustrate the impact of interaction rules on distributed computer control systems. Three basic classes of simulation were carried out and are described in more detail below (sections 4.1, 4.2 and 4.3).

Figures 1, 2 and 3 show the time history for the distributed system. Each plot consists of two subplots — one for each processing node in the distributed system. The plots show the network traffic generated by the distributed system and processor loading. The processor load trace indicates the % time during the interval that the processor was not idle, this includes all processing activities not just those related to the distributed system. In both cases the average over an interval (indicated on the subplot x-axis label) is plotted. Notice that the zero second points for corresponding simulation plots do not necessarily match (there is typically about a 2-4 second mismatch). For the purpose of this discussion the mismatch is not important as the plots are simply used to demonstrate the system operation.

4.1 Static operation

The simulation agent was started on a node in the distributed system with the error signal class set to full PID (i.e., the controller structure required is PID). The evolution of the distributed controller was then recorded. The simulations demonstrate the establishing of a distributed controller and "steady state" operation of the controller.

Figure 1 shows an increase in processor loading on Node B at time zero corresponding to the control signal agent launch, shortly there-after a decision agent is launched on Node A, indicated by the peak in processor loading. Coinciding with the agent launches is an increase in network traffic which peaks (agent launch) and then retains small periodic peaks (inter-agent communication). At around 75 seconds processor loading on both nodes peaks again, the decision agent has identified a structure causing a differential agent to launch on Node A and integral agent on Node B. After the agent launches both the processor loading and network traffic visibly increase a result of the increased controller complexity. At 175 seconds the loading on Node A peaks, this is combination of user input (first peak) stopping the simulation and agents being removed as a result of the halted simulation. The corresponding peak on Node B is due solely to agents being removed. After the loading and traffic peaks all is quiet, only the stopped simulation agent remains in the distributed system.

Similar simulations (i.e., all requiring full PID, but no restriction on where agents are launched) showed that the typical maximum and average processor loading is 42.5% and 5.5% respectively (user induced loading was omitted for these calcu-

lations). Similarly the typical maximum and average network traffic was recorded at 740 bytes/sec and 105 bytes/sec respectively.

4.2 Dynamic operation

In this case the simulation was started with a proportional error signal class, allowed to reach a steady state when the error signal class was changed to full PID. The simulations are used to illustrate the ability of the interaction rule approach to adapt to plant changes and demonstrate the use of operational knowledge in changing distributed system structure.

Figure 2 shows the simulation results. From zero to 25 seconds the peak in loading and traffic indicates the launch of both the control signal and decision agents on Node B. The large loading peak on Node A at 40 seconds corresponds to the user changing the error signal class from proportional only to full PID. The resulting effect is evident at 75 seconds where a differential agent is launched on Node A and an integral agent on Node B. After the additional agents are launched both the traffic and processor load increases as expected, the simulation is stopped around 160 seconds.

Using several simulations and comparing the load and traffic averages for the proportional only and full PID cases indicates that the processor load changes marginally by about 1% while the network traffic changes significantly from 105 bytes/sec to 158 bytes/sec.

4.3 Fault conditions

The final simulation case had the simulation started with a differential error signal class which was allowed to reach steady state. The control signal agent was then forcibly (i.e., user intervention) removed from the distributed system in order to mimic a fault condition and the distributed system left to reach a steady state yet again. These simulations demonstrate the ability for interaction rules to provide a degree of fault tolerance in distributed systems.

Figure 3 shows the distributed system history for a fault simulation. Initially the control signal and decision agents are launched on Node A at 50 seconds the differential agent is also launched on Node A. Notice that there is no network traffic during this period and that the Node B node is idle. After 100 seconds the user removes the control signal agent from the distributed system and as a result the decision and differential agents are automatically removed. Shortly thereafter (i.e, 1-2 seconds) a control signal agent is launched on

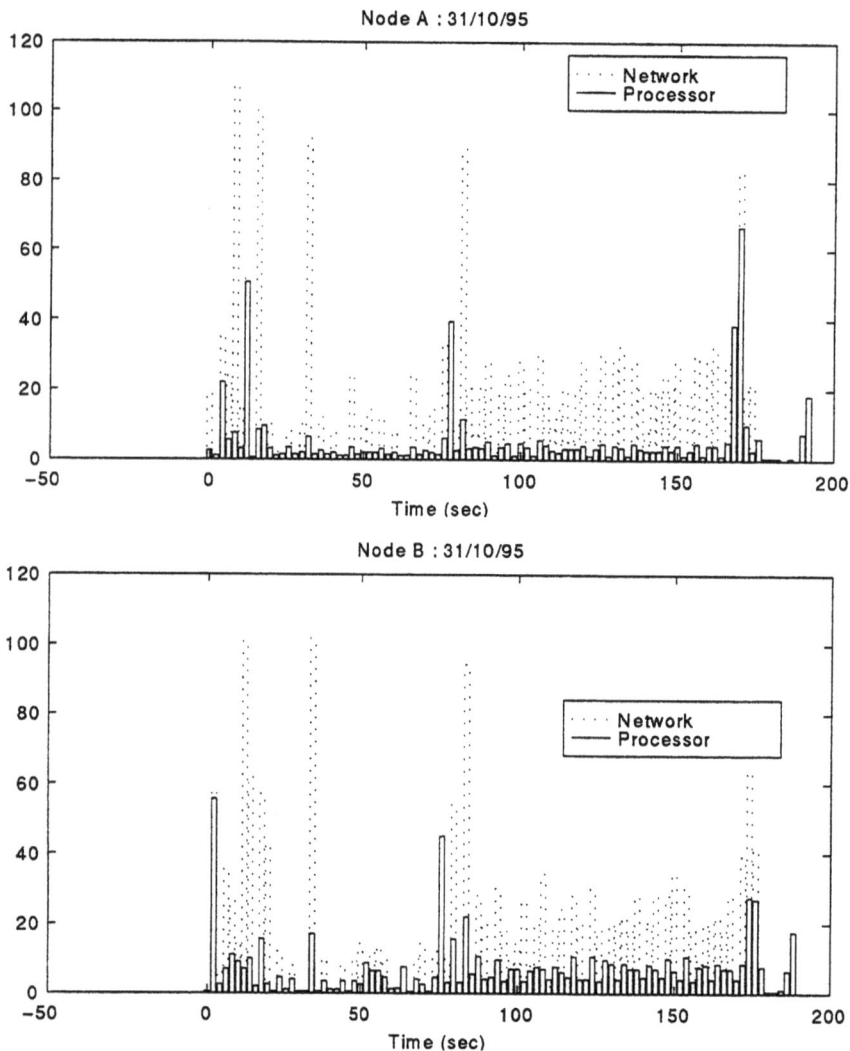

Fig. 1. Simulated static operation for measurement intervals of 2 seconds. Network load is measured in units of 10 Bytes per second and processor loading as % time active in interval.

Node B by the simulation agent, the control signal agent then launches a decision agent, also on Node B. By 150 seconds a differential agent is launched — coincidently also on Node B.

4.4 Comments

During testing additional simulations were carried out with two independent simulation agents in the distributed system each of which resulted in the establishment of independent control signal, decision, integral and differential agents. Also during development the control signal agent template occasionally produced a memory fault (subsequently corrected) causing the active agent to crash and remove connected decision and/or differential, integral agents, however the fault tolerant characteristic took over and restarted a new control signal agent which then reconstructed the lost distributed system structure.

During development one concern was that automatically launched agents would not be evenly distributed across the distributed system. The simulation runs indicated that 49% of the agents were launched on Node B and 51% on Node A. Of more interest was the difference in the number of agents launched on each node during a simulation run (the agent mismatch). An average agent mismatch of 1.5 was recorded, which is a little high (ideal would be less than 1) but is due in some extent to a group of simulations which required an odd number of agents. The mismatch could be reduced by changing the random method of launching agents to a more deterministic one which includes information on where agents were previously launched.

Another concern was the effect of dynamically launching agents on both the network and processor performance. The simulations showed that under a worst case the network load peaked at around 1000 bytes/sec. For an ethernet connection capable of end-to-end data transfer (i.e., excluding protocol overheads) in the order of "... tens of kilobits (rather than megabits)" (Halsall 1988)

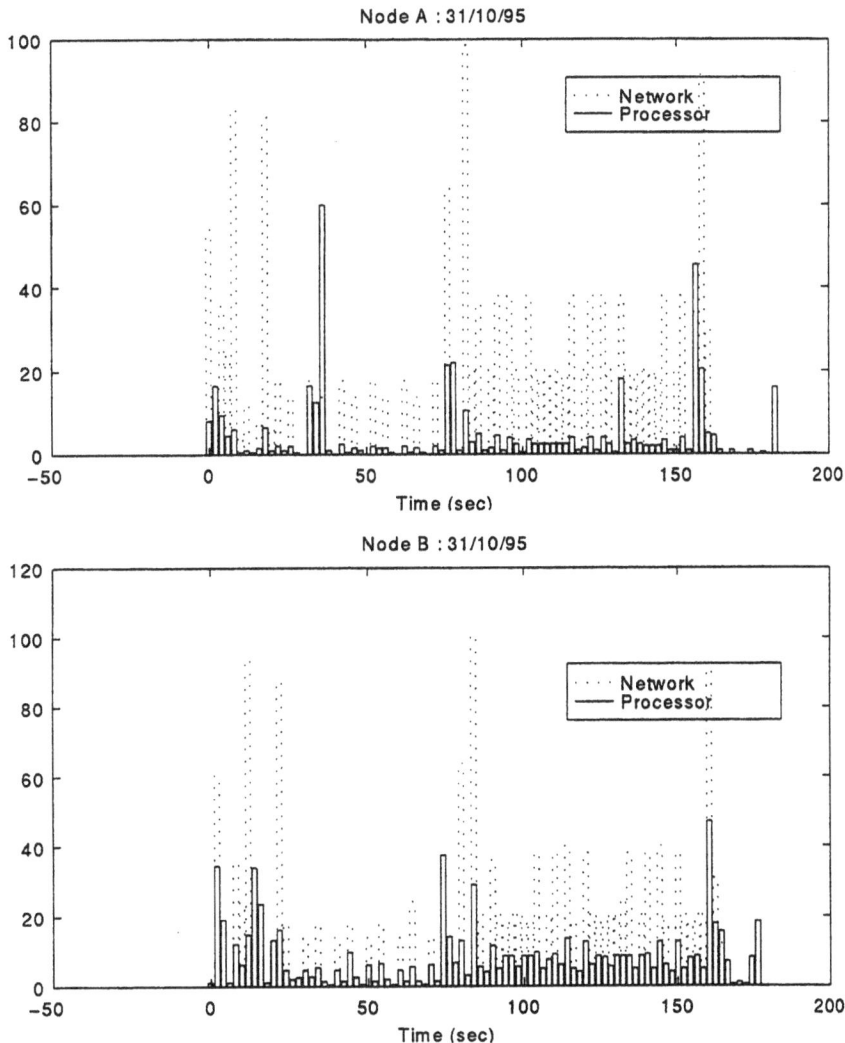

Fig. 2. Simulated dynamic operation for measurement intervals of 2 seconds. Network load is measured in units of 10 Bytes per second and processor loading as % time active in interval.

the increased network loading is minimal[2]. Worst case processor loading peaked at around 50%, again this is acceptable considering that a user can drive the load to 60-70% simply by mouse movement! Note that the worst case for both the traffic and processor load was never sustained - indicating that initial performance concerns were largely vindicated.

A final point worth noting is that the simulations indicate that a "solution" (read agent configuration) is built up or evolves around the initial simulation agent. At stages during the building process a partial "solution" is present. The availability of partial solutions and the tendency for the distributed system to evolve indicate progressive reasoning which is important for real-time systems (MacLeod and Lun 1992). For example during the static operation simulations described in section 4.1 the distributed system is able to

supply a proportional control signal long before the correct or optimal control signal class has been identified. This progressive reasoning is a by-product of the design approach adopted which focusses on system behaviour and is implemented via interaction rules.

5. CONCLUSION

Distributed control systems focus on the use of autonomous agents and how they cooperate and interact to achieve a system-wide goal. The performance of a distributed system design philosophy which distinguishes between a priori and operational knowledge and describes how dynamic structuring can be achieved by using the ideas of agent templates and interaction rules has been discussed.

The effect of agent templates and interaction rules on a distributed system is highlighted through an implemented distributed controller that simulates

---

[2] Network analysers connected to the departmental Ethernet indicate that network saturation occurs at around 40% loading, i.e., 4 Mbits/s.

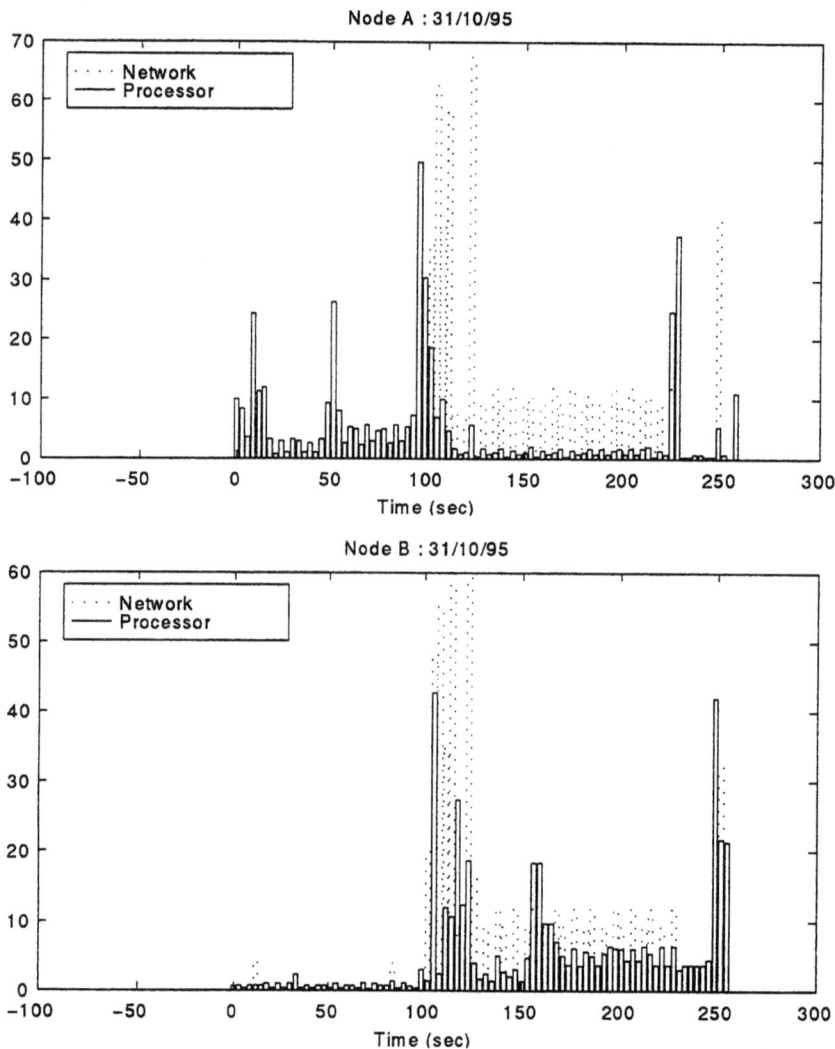

Fig. 3. Simulated fault condition for measurement intervals of 2 seconds. Network load is measured in units of 10 Bytes per second and processor loading as % time active in interval.

a simple conventional controller. In particular, results highlight the ability of the distributed system to

- adapt as operating conditions/requirements change.
- recover from fault conditions.
- provide partial solutions and perform progressive reasoning.

Through simulation the effect of this approach to distributed control system design on the distributed system, both in terms of network traffic and processor loading, is shown to be minimal.

6. REFERENCES

Bird, S. (1993). Toward a taxonomy of multi-agent systems. *International Journal of Man Machine Studies* **39**(4), 689–704.

Cohen, J. (1991). A Model of Process Interaction in Real-Time Distributed Computer Control Systems. PhD thesis. University of the Witwatersrand.

Davis, R. and R. G. Smith (1983). Negotiation as a metaphor for distributed problem solving.. *Artificial Intelligence* **20**, 63–109.

Decker, K. (1987). Distributed problem-solving techniques: A survey. *IEEE Transactions On Systems, Man, And Cybernetics* **17**(5), 729–740.

Durfee, E. (1991). The distributed artificial intelligence melting pot. *Transactions on Systems Man and Cybernetics* **21**(6), 1301–1306.

Franklin, G., J. Powell and A. Emami-Naeini (1986). *Feedback Control of Dynamic Systems*. Chap. 3, pp. 88–117. Addison-Wesley.

Halsall, F. (1988). *Data Communications, Computer Networks and OSI*. Chap. 8, pp. 373–376. 2nd ed.. Addison-Wesley.

Lesser, V. and D. Corkill (1981). Functionall acurate, co-operative distributed systems. *IEEE Transactions on System Man and Cybernetics* **11**(1), 81–96.

Lueth, T., T. Laengle and J. Heinzman (1995). Dynamic task mapping for real-time controller of distributed cooperative robot systems. In: *13th IFAC Workshop on Distributed Computer Control Systems*. IFAC. Toulouse-Blagnac (France). pp. 37–42.

MacLeod, I. and A. Stothert (1994). A simulation study of distributed intelligent control for a deep shaft mine winder. In: *Proc. IFAC Symposium on Artificial Intelligence in Real-Time Control, Valencia*. pp. 459–464.

MacLeod, I. and V. Lun (1992). Progressive reasoning for real-time intelligent computing. *IEEE Control Systems* **12**(2), 79–83.

Stothert, A. and I. MacLeod (1995). Research issues in the dynamic structuring of distributed intelligent controllers. In: *IFAC Youth Automation Conference*. IFAC. Beijing (China). pp. 608–612.

Stothert, A. and I. MacLeod (1997). Using intelligent agent templates for dynamic structuring of distributed computer control systems. *Engineering Applications of Artificial Intelligence*. In press.

Sycara, K. (1989). Multiagent compromise via negotiation. In: *Distributed Artificial Intelligence Volume II* (L. Gasser and M. Huhns, Eds.). Chap. 6, pp. 119–138. Morgan Kaufmann.

Werner, E. (1989). Cooperating agents: A unified theory of communication and social structure. In: *Distributed Artificial Intelligence Volume II* (L. Gasser and M. Huhns, Eds.). Chap. 1, pp. 3–36. Morgan Kaufmann.

ADAPTIVE GENERALIZED PREDICTIVE CONTROL ALGORITHM IMPLEMENTED OVER AN HETEROGENEOUS PARALLEL ARCHITECTURE

H. A. Daniel*, A. E. B. Ruano**

*, ** Unidade de Ciências Exactas e Humanas, Universidade do Algarve
** Institute of Systems and Robotics, Portugal
Email:* hdaniel@beethoven.si.ualg.pt, ** aruano@mozart.si.ualg.pt
Tel: * +351 89 800950   ** +351 89 800912

Abstract: In this paper a parallel implementation of an Adaptive Generalized Predictive Control (AGPC) algorithm is presented. Since the AGPC algorithm needs to be fed with knowledge of the plant transfer function, the parallelization of a standard Recursive Least Squares (RLS) estimator and a GPC predictor is discussed here. Also, since a matrix inversion operation is required in the GPC predictor algorithm, special attention is given to its parallelization. A small DSP network with up to 3 processors is used to investigate, the performance of the parallel implementation. To exploit an heterogeneous architecture the parallel algorithm is mapped over a network builded up of transputers as communication elements, and DSPs as computing elements. Further some heterogeneous topologies are compared. Execution times and efficiency results of the RLS and GPC steps are presented to show the performance of the parallel algorithm, over different topologies.

Keywords: Heterogeneous parallel architectures, Parallel algorithms, Adaptive control, Multiprocessing Systems, Digital signal processors

1. INTRODUCTION

The GPC algorithm was introduced in the past decade (Clarke, et al., 1987a, 1987b) and has subsequently been proved to be superior to other self-tuning algorithms such as the ones based on pole-placement and generalized minimum variance. In fact the GPC algorithm conserves its robustness even when the plant dead-time and model order are unknown (Aström and Wittenmark, 1989), but the cost of achieving such robustness is a substantial cost in computational overhead. Employing the new powerful generation of parallel processors, such as DSPs and transputers, its implementation can be speeded up, therefore making it a valuable proposition for the control of plants with tighter specifications in terms of sampling time. As the adaptive GPC algorithm needs to know the plant transfer function, a first stage in the algorithm must be an RLS estimator. The estimated parameters are then injected in the predictor which computes the predicted plant output and the control signal with a subsequent minimisation of a suitable loss function.

The block diagram of this algorithm is presented in figure 1.

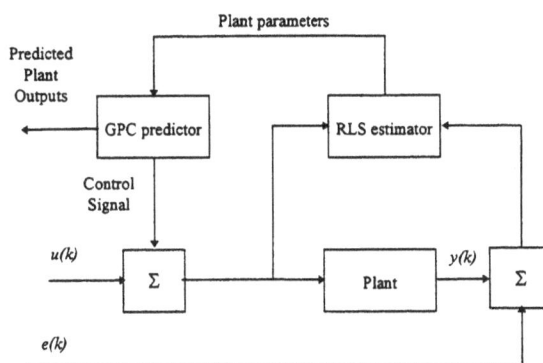

Fig. 1. Block diagram of a plant controlled by a GPC predictor

2. RECURSIVE LEAST SQUARES ALGORITHM

Let the plant model be a CARIMA representation:

$$A(q^{-1})y(k) = q^{-d}B(q^{-1})u(k-1) + \frac{C(q^{-1})}{\Delta}e(k) \tag{1}$$

where $y(k)$ is the plant output at time step k, $u(k-1)$ is the plant input at time step $k-1$, $e(k)$ is an uncorrelated random sequence, Δ is the difference operator $1 - q^{-1}$, d is the time delay and $A(q^{-1})$, $B(q^{-1})$, $C(q^{-1})$ are polynomials in the backward shift operator q^{-1}:

$$A(q^{-1}) = 1 + a_1 q^{-1} + ... + a_{na} q^{-na}$$
$$B(q^{-1}) = b_0 + b_1 q^{-1} + ... + b_{nb} q^{-nb}$$
$$C(q^{-1}) = 1 + c_1 q^{-1} + ... + c_{nc} q^{-nc}$$

Let the following vectors be the unknown parameters vector and the regression vector or data vector, respectively:

$$\theta^T = \left[-a_1,...,-a_{n_a}.b_0,...,b_{n_b},c_1,...,c_{n_c}\right]$$
$$\mathbf{x}^T(k) = \left[y(k-1),...,y(k-n_a),u(k-1),...\right.$$
$$\left....,u(k-n_b-1),e(k-1),...,e(k-n_c)\right]$$

and let $\mathbf{P}(k)$ be the covariance matrix, initialised as the Identity matrix. The following set of equations defines a standard RLS estimator, optimised in terms of computational operations:

$$\varepsilon(k) = y(k) - \mathbf{x}^T(k)\theta(k-1) \tag{2}$$

$$\mathbf{K}_b(k) = \mathbf{P}(k-1)\mathbf{x}(k) \tag{3}$$

$$\mathbf{K}(k) = \frac{\mathbf{K}_b(k)}{\lambda + \mathbf{x}^T(k)\mathbf{K}_b(k)} \tag{4}$$

$$\mathbf{P}(k) = \frac{\mathbf{P}(k-1) - \mathbf{K}(k)\mathbf{K}_b(k)^T}{\lambda} \tag{5}$$

$$\theta(k) = \theta(k-1) + \mathbf{K}(k)\varepsilon(k) \tag{6}$$

3. GPC PREDICTOR

The GPC predictor minimisation criteria is given by:

$$J(N_1, N_2, N_u) = E\left[\sum_{j=N_1}^{N_2} \delta(j)\left[\hat{y}(k+j \mid k) - w(k+j)\right]^2 + \cdots\right.$$
$$\left.\cdots + \sum_{j=1}^{N_u} \lambda(j)\left[\Delta u(k+j-1)\right]^2\right] \tag{7}$$

where $E[...]$ is the expectation operator, $\hat{y}(k+j \mid k)$ is an optimal j steps ahead predictor using data available until time step k, N_1 is the minimum cost horizon, N_2 is the maximum cost horizon, Nu is the control horizon, $\lambda(j)$ and $\delta(j)$ are weighting sequences (($\lambda(j)$ is considered to be constant and $\delta(j) = 1$), and $w(k+j)$ and is the future reference trajectory. To obtain the output j steps ahead, we need to solve the diophantine equation:

$$C(q^{-1}) = E_j(q^{-1})\tilde{A}(q^{-1}) + q^{-j}F_j(q^{-1}) \tag{8}$$

where $\tilde{A}(q^{-1}) = \Delta A(q^{-1})$ with deg $E_j = j - 1$, deg $F_j = n_a$. $E_j(q^{-1})$ and $F_j(q^{-1})$ can be obtained dividing $C(q^{-1})$ by $\tilde{A}(q^{-1})$ until the remainder can be factored as $q^j F_j(q^{-1})$, where $E_j(q^{-1})$ is the quotient (see Camacho and Bordons, 1994), or recursively (Clarke et al., 1987; Camacho and Bordons, 1994)

Considering now the following sequence of j steps ahead optimal predictions:

$$\hat{y}(k+d+1 \mid k) = G_{d+1}(q^{-1})\Delta u(k) + F_{d+1}(q^{-1})y(k)$$
$$\hat{y}(k+d+2 \mid k) = G_{d+2}(q^{-1})\Delta u(k+1) + F_{d+2}(q^{-1})y(k)$$
$$\vdots$$
$$\hat{y}(k+d+N \mid k) = G_{d+N}(q^{-1})\Delta u(k+N-1) + F_{d+N}(q^{-1})y(k)$$

where G is obtained with:

$$G_j = E_j.B, \qquad j = 1,...,N \tag{9}$$

which can be written as (see Aström and Wittenmark, 1989)

$$\mathbf{y} = \mathbf{G}\Delta\mathbf{u} + \mathbf{f} \tag{10}$$

where:

$$\Delta\mathbf{u}^T = \left[\Delta u(k), \Delta u(k+1), \cdots, \Delta u(k+N-1)\right]$$

$$\mathbf{y}^T = \left[\hat{y}(k+d+1 \mid k), \hat{y}(k+d+2 \mid k), \cdots, \hat{y}(k+d+N \mid k)\right]$$

$$\mathbf{G} = \begin{bmatrix} g_0 & 0 & \cdots & 0 \\ g_1 & g_0 & \cdots & 0 \\ \vdots & \vdots & \ddots & \vdots \\ g_{N-1} & g_{N-2} & \cdots & g_0 \end{bmatrix}, \begin{array}{l} g_n = n^{th}\text{coefficient} \\ \text{of } G_{d+N} \\ n = 0,...,N-1 \end{array}$$

$$\mathbf{f} = \begin{bmatrix} (G_{d+1}(q^{-1})-g_0)\Delta u(k) + F_{d+1}(q^{-1})y(k) \\ (G_{d+2}(q^{-1})-g_0-g_1 q^{-1})\Delta u(k) + F_{d+2}(q^{-1})y(k) \\ \vdots \\ (G_{d+N}(q^{-1})-g_0-g_1(q^{-1}-\cdots-g_{N-1}(q^{-(N-1)}))\Delta u(k) + F_{d+N}(q^{-1})y(k) \end{bmatrix}$$

$$N = N_2 - N_1 + 1 = N_u$$

$\Delta\mathbf{u}$, the minimum solution of (7), can be computed as (see Aström and Wittenmark, 1989):

$$\Delta\mathbf{u} = (\mathbf{G}^T\mathbf{G} + \lambda\mathbf{I})^{-1}\mathbf{G}^T(\mathbf{w} - \mathbf{f}) \tag{11}$$

where:

$$\mathbf{w} = \left[w(k+d+1), w(k+d+2), \cdots, w(k+d+N)\right]$$

4. PARALLEL ALGORITHM

4.1 Architectures

The bottom level of the network consists in a strategy of routing through a 3 workers network. The physical connections of an homogeneous network can be seen on Fig. 2. The routing strategy assures that a degenerated network is still a valid environment for broadcasting.

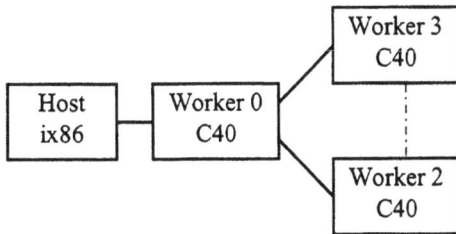

Fig.2. Homogeneous Network of 3 workers

Two heterogeneous networks builded of transputers and DSPs are presented on Fig. 3. Note that in network a) all the communications are handled by transputers while in b) the transputer only provides an interface between the Host and the DSPs. In both figures a dashed line represents a link that, when used, establishes a ring topology. When used this link can also provide point to point communication between every node in the network. The message broadcasting is transparent to the upper level, and whatever the topology used, there are two communication primitives:

> *send*   *(destination, sender, data, length)*
> *receive* *(destination, sender, data, length)*

However sometimes, for sake of clarity, this primitives are expressed as:

> *send*   *(destination, data₁, data₂, ... dataₙ)*
> *receive* *(destination, data₁, data₂, ... dataₙ)*

where destination can have 3 values when sending the message: ROOT- send message to root; TO_ALL - broadcast message to all workers except to the one who sent it; and WRKXX - send message to processor XX, $1<XX<$ N. of workers. When receiving a message from the network, information of who sent it and to whom is destinated can be obtained from *destination* and *sender*. To allow this transparency, each node incorporates 2 routing high-priority processes. One handles incoming messages while the other handles the outcoming ones. Fig. 4 shows a typical node. The communication section consists in two routers. When router **in** detects a message, it holds it in a buffer until the computing

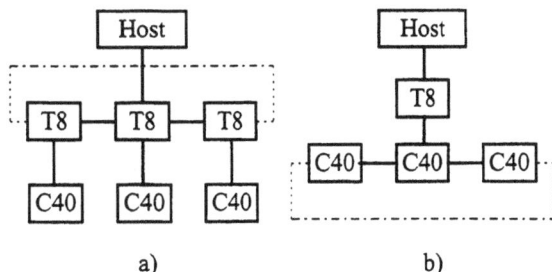

Fig.3. Heterogeneous Networks

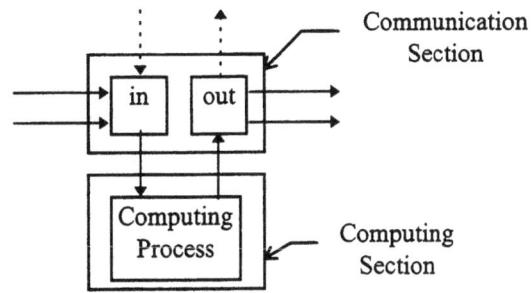

Fig.4. Typical node

process requests it. The **out** router receives a *send* primitive and, depending of the value of *destination*, sends it to the corresponding out links. The computing section can be seen as the node floating point unit (FPU).

For an homogeneous architecture both sections represented in Fig. 4 reside in the same processor. Notice that for a network such as the one in Fig. 2, worker 0 communication section must use the dashed links, so that communication with the host may be provided. For an heterogeneous architecture, such as the one shown in Fig. 3 a), the communication section is mapped on the transputers and the computing section on the DSPs. Again the transputer connected with the host must use the dashed links on Fig. 4. For the network presented in fig. 3 b) both sections must be mapped on the DSPs. In this network all nodes must use the dashed links to communicate with the host via the T8.

4.2 Introduction to Matricial Operations

Considering now the parallelization of the AGPC algorithm, a standard matrix partitioning can be employed so that the data is divided by the available processors executing the same code. For instance if one would want to compute the product of two vectors **A** and **B**:

$$c = \mathbf{A} \cdot \mathbf{B}$$

$$c = \begin{bmatrix} a_{11} & a_{12} & a_{13} & \cdots & a_{1m} \end{bmatrix} \cdot \begin{bmatrix} b_{11} \\ b_{21} \\ b_{31} \\ \vdots \\ b_{m1} \end{bmatrix}$$

in two processors, the partition can be made as follows:

$$c = \begin{bmatrix} \mathbf{A}_1 & \vdots & \mathbf{A}_2 \end{bmatrix} \cdot \begin{bmatrix} \mathbf{B}_1 \\ \cdots \\ \mathbf{B}_2 \end{bmatrix}$$

where:

$$\mathbf{A}_1 = \begin{bmatrix} a_{11} & a_{12} & \cdots & a_{1\,(m/2-1)} \end{bmatrix}$$

$$\mathbf{A}_2 = \begin{bmatrix} a_{1\,(m/2)} & a_{1\,(m/2+1)} & \cdots & a_{1\,m} \end{bmatrix}$$

$$\mathbf{B}_1{}^T = \begin{bmatrix} b_{11} & b_{21} & \cdots & b_{(m/2-1)\,1} \end{bmatrix}$$

$$\mathbf{B}_2{}^T = \begin{bmatrix} b_{(m/2)\,1} & b_{(m/2+1)\,1} & \cdots & b_{m\,1} \end{bmatrix}$$

allocating vectors $\mathbf{A}_1$ and $\mathbf{B}_1$ in one processor and vectors $\mathbf{A}_2$ and $\mathbf{B}_2$ in the other then $c = c_1 + c_2$, where $c_1 = \mathbf{A}_1 . \mathbf{B}_1$ and $c_2 = \mathbf{A}_2 . \mathbf{B}_2$ as shown in Fig. 5.

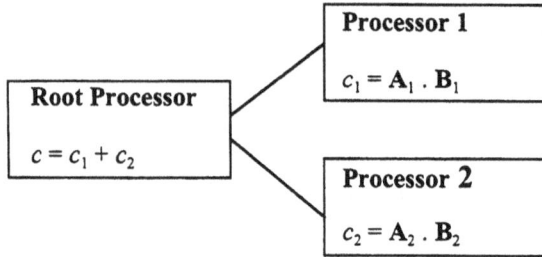

Fig. 5. Parallel computation of vectorial product using 2 processors

A similar kind of partition could be used in matrix computation. Lets suppose one want to multiply two square matrices with two processors:

$$\mathbf{C'} = \mathbf{A'} . \mathbf{B}$$

Allocating the top half partition of matrix $\mathbf{A}$ in one processor, the bottom one in the other and the complete matrix $\mathbf{B}$ in both processors, a standard matrix multiplication algorithm will output the top partition of the resultant matrix $\mathbf{C}$ in one processor and the bottom partition in the other one. From the union of both parts the complete $\mathbf{C}$ matrix is obtained. The same method could be used to multiply a matrix by a vector. Note that the quotation mark following the variable name indicates that the variable is a partition of the full matrix. Since it is assumed that the code is identical for every worker, when referring to a partition of a matrix it is assumed that for different workers the allocated partition is different.

The matrix addition is then trivial, since equal half parts of both matrices are allocated in the same processor. An element by element addition is then performed in both processors and, as in the matrix multiplication, the result matrix is obtained from the union of both partial results. This concept is applied to the parallelization of the algorithm.

4.3 Parallel Algorithm

The simplified parallel algorithm is presented in 2 columns. The left one represents the code running in the Root processor, which is assumed to be the processor that is physically connected to the host for every architecture, in parallel with the code running in every worker, which is represented in the right column. The algorithm is expressed mainly in structured English and standard math notation. The introduced notation was described in the above section. Note that in this level the only concern with communications is where to and how much information is sent. So the estimator can be parallelized as follows:

First set $\mathbf{P}(k-1) = \mathbf{I} * 10^6$; $\mathbf{x}(k-1)$ to a null vector. Let Z be the number of workers in the mesh. For each time step k do:

1) Update data vector $\mathbf{x}(k)$, and compute equation 2.

Root	Worker
receive $y(k)$, $u(k)$ from HOST;	
send (TO_ALL, $y(k)$, $u(k)$)	receive ($y(k)$, $u(k)$)
	actualise $\mathbf{x}(k)$
	$\mathbf{T} = \mathbf{x'}(k)^T \theta'(k-1)$
	send (TO_ALL, $\mathbf{T}$)
	for j=0 to Z-1 receive ($\mathbf{T}(j)$)
	$\varepsilon(k) = \mathbf{y}(k) - \displaystyle\sum_{j=0}^{Z-1} \mathbf{T}(j)$
send (TO_ALL, $\varepsilon(k)$)	receive ($\varepsilon(k)$)

2) Compute equation 3.

Root	Worker
	$\mathbf{K}_b{}'(k) = \mathbf{P'}(k-1)\, \mathbf{x}(k)$
	send (TO_ALL, $\mathbf{K}_b{}'(k)$)
	receive $\mathbf{K}_b(k)'$ partitions from every worker; form $\mathbf{K}_b(k)$;

3) Compute equation 4.

Root	Worker
	$\mathbf{T} = \mathbf{x'}(k)^T \mathbf{K}_b{}'(k)$
	send (TO_ALL, $\mathbf{T}$)
	for j=0 to Z-1 receive ($\mathbf{T}(j)$)
	$den = \lambda + \displaystyle\sum_{j=0}^{Z-1} \mathbf{T}(j)$
	$\mathbf{K'}(k) = \dfrac{\mathbf{K}_b{}'(k)}{den}$

4) Compute equation 5

Root	Worker
	$\mathbf{P'}(k) = \dfrac{\mathbf{P'}(k-1) - \mathbf{K'}(k)\mathbf{K}_b{}'(k)^T}{\lambda}$

5) Compute equation 6, and send the parameter vector $\theta(k)$ to the Host.

Root	Worker
	$\theta'(k) = \theta'(k-1)\ \mathbf{K'}(k)\ \varepsilon(k)$
receive $\theta'(k)$ partitions from every worker; form $\theta(k)$;	send (TO_ROOT $\theta'k$))
send (TO_ALL $\theta'(k)$)	receive ($\theta(k)$)
send $\theta(k)$ to HOST;	

A similar scheme can be used to specify the predictor parallelization. For each time step k do:

1) Update future set point sequence $w(k+N)$, solve diophantine identity, and form G, $\mathbf{G}$, $\mathbf{f}$.

Root	Worker
receive $w(k+N)$ from HOST;	
send (TO_ALL, $w(k+N)$)	receive ($w(k+N)$)
	actualise **w'**
	$C = E_j(q^{-1})\, \tilde{A}(q^{-1}) + q^{-j} F_j(q^{-1})$
	form **G'**, **G'**, **f'**;
	send (TO_ALL, **G'**)
	receive **G'** partitions from every worker; form **G**;

2) Compute equation 11.

Root	Worker
	buffer' = **G'^{T}** . **G** + λ**I**
	inverse' = (**buffer^{-1}**)'
	Δ**u'** = **inverse'** . **G'^{T}** . (**w'** - **f'**)
	send (TO_ALL, Δ**u'**)
	receive Δ**u'** partitions from every worker; form Δ**u**;

The only drawback here is the matrix inversion. This is handled by a parallel implementation of the Gauss-Jordan elimination method (Daniel and Baltazar, 1994). Consider the following equations:

$$A_{kj}^{(k+1)} = \frac{A_{ii}^{(k)}}{A_{kk}^{(k)}} \qquad (12)$$

$$A_{ij}^{(k+1)} = A_{ij}^{(k)} - \frac{A_{ik}^{(k)}}{A_{kk}^{(k)}} A_{kj}^{(k)} \qquad (13)$$

where: A is a n x 2n matrix
i = 1, ..., k-1, k+1, ..., n
j = k+1, ..., n+1
k = 1, 2, ... n
$A_{kk}^{(k)} = pivot$

Let **A'** = [**buffer'** | **I'**], where **I** is the identity matrix.

k = 1, 2, ..., n

    i) The worker who contains line **k** of **A** solves equation 12 over that line and sends the result to every other processor
    ii) Equation 13 is applied to every line of **A'**, except line **k**, in every processor.

at step n, in **A'** is obtained:

$$\mathbf{A'} = [\; \mathbf{I'} \mid (\mathbf{buffer}^{-1})' \;]$$

3) Compute the predicted outputs **y** and send it to the host (equation 10).

Root	Worker
	y' = **G'** . Δ**u** + **f'**
receive **y'** partitions from every worker; form **y**;	send (TO_ROOT, **y'**)
send **y'** to HOST;	

4) Compute the control signal $u(k)$
To compute the control signal the following equation must be solved:

$$u(k) = u(k-1) + \Delta\mathbf{u}(1) \qquad (13)$$

Since only the first line of **G** is needed the processor which holds it should execute the following piece of code:

Root	Worker
	$u(k) = u(k - 1) + \Delta\mathbf{u}(1)$
receive ($u(k)$)	send (TO_ROOT, $u(k)$)
send $u(k)$ to HOST;	

5. RESULTS

Results of parallel execution times and efficiencies for each major step of the algorithm, and for the complete parallel adaptive GPC algorithm were measured over a network composed of T805 - 25 MHz and TMS320C40 - 50 MHz. Fig. 6 and Fig. 7 shows the performance comparation of the total parallel algorithm mapped over the architectures presented in Fig. 2 (Net1), Fig. 3 a) (Net2) and Fig. 3 b) (Net3) with a homogeneous transputer network (Net0). The number of workers in the network is indicated after the network mnemonic. The systems used to test the algorithm were two CARIMA plants with coloured noise. The smallest was a 2nd, order system with 5 parameters, m. The other one was a 5th order system with m = 15. A predictive horizon, N, of up to 100 steps was used.

As can be seen from Fig. 6, where the performance of the slowest architectures, Net2 and Net0, are shown, Net2 with one node only is much faster than Net0 with three. This is because the TMS320C40 DSP has a much faster FPU than the T805 Transputer. Net2 uses transputers to handle communications only and C40s as FPUs. Fig.7 shows the performance of the fastest architectures implemented, Net1 and Net3. Note that for the case of one node both architectures have the same performance. In this case Net2 with two and three nodes are faster than this architectures. However if one increases the nodes of these last architectures they prove to be much faster, with Net3 being the fastest. This is explained by the allocation of the MASTER task. In Net1, the homogeneous C40 architecture, the MASTER task is allocated in worker 0. So inspite of the small computing effort that this task requires, as it only handles communications with

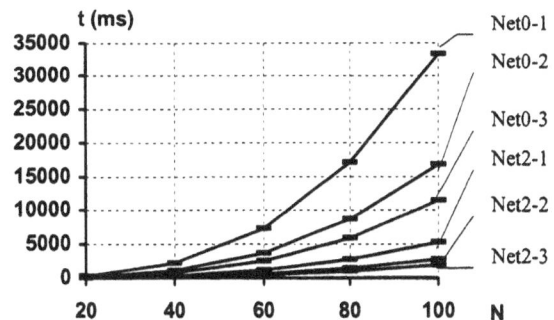

Fig. 6. - Execution times of architectures Net0 and Net2 with m = 5 and m = 15

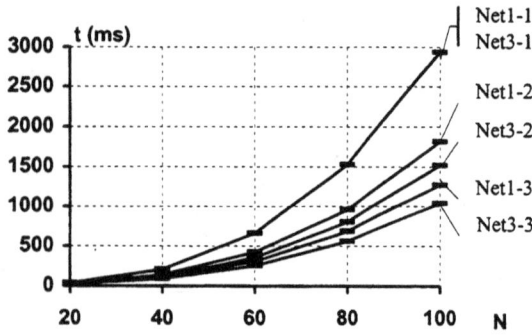

Fig. 7. - Execution times of architectures Net1 and
Net3 with m = 5 and m = 15

the host, still it slows down worker 0. However in
Net3 this task is allocated in the transputer so
releasing worker 0 of this additional effort.

Comparing the fastest and the slowest networks it can
be seen that, as N increases, there is also an
increasing ratio in the performance of the two
networks. For a 1 worker only network, with N = 20
and m = 15, Net3 is 8.6 times faster than Net0.
However increasing N to 100 the ratio is 11.4. Note
that for m = 5 and m = 15 the difference in terms of
computational time is not meaningful, since the
estimator step is far less complex than the predictor.
So the GPC step is the time consuming one. This
forces AGPC efficiency to follow GPC, as N
increases, since the RLS step efficiency only depends
of m, as can be seen on Figs. 8 through 10.

The efficiency of the estimator in every case is higher
with 2 workers, as can be seen on Fig. 8. Looking at
Fig. 10 it can be seen that the higher efficiency is
achieved by Net0, however this one is the slowest of
the architectures implemented. Net2 is the one that
follow Net0 the best, however for long range
horizons, Net3, the fastest network, almost reaches
Net0 efficiency. Net1 is the less efficient architecture.

6. CONCLUSIONS

The architectures implemented are very efficient for
long range predictive controllers. The number of
parameters to estimate in the plant almost does not
affect the AGPC efficiency. Although the
architectures that use transputers as communication
elements have proved to be the most efficient ones,

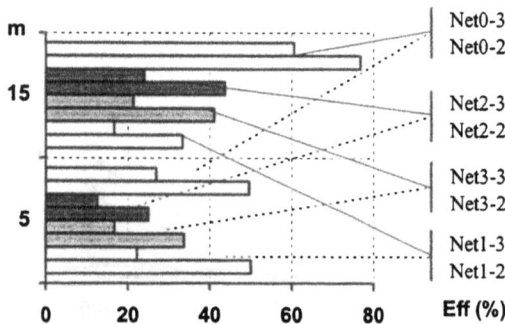

Fig. 8. - Efficiency of the RLS step for 2 and 3
workers networks, with m = 5 and m = 15.

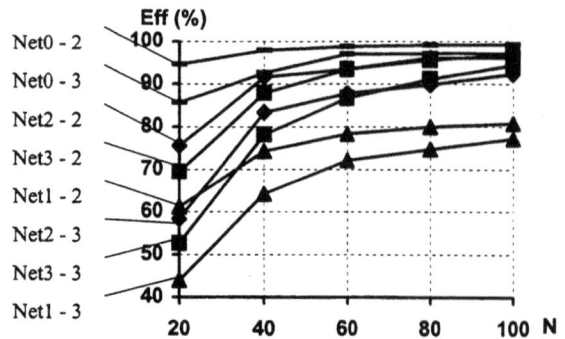

Fig. 9. - Efficiency of the GPC step for 2 and 3
workers network, with m = 15.

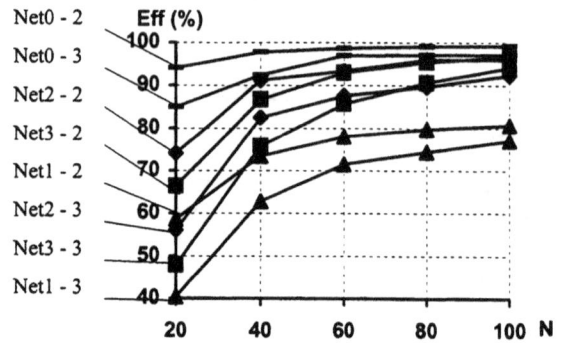

Fig. 10. - Efficiency of the AGPC parallel algorithm
for 2 and 3 workers network, with m = 15.

the fastest architecture is the one that uses C40s as
computation elements and a transputer to interface
with the host, Net3. This architecture is also the most
efficient of the architectures that uses C40s as
computational elements for long range predictive
horizons. However, for small horizons, its efficiency
is just above the one with the smallest efficiency -
Net1. Future work will intend to raise the efficiency
of the architectures based on C40s for small
predictive horizons, without loss of execution time.
Also faster computational routines will be addressed.

ACKNOLEDGEMENTS

Helder Daniel acknowledges the financial support of
"Programa Praxis XXI" for this work

REFERENCES

Aström, K. J., and B. Wittenmark (1989). *Adaptive
Control*. Addison - Wesley.
Camacho, E. F. and J. C. Bordons (1994). *Model
Predictive Control in the Process Industry*. Springer
Verlag
Clarke, D. W., C. Mohtadi and P. S. Tuffs (1987a)
Generalized Predictive Control - Part I. The Basic
Algorithm. In: *Automatica*, **23**, 137 - 148
Clarke, D. W., C. Mohtadi and P. S. Tuffs (1987b).
Generalized Predictive Control - Part II. Extensions
and Interpretations. *Automatica*, **23**, 149 - 160
Daniel, H. and S. Baltazar, S. (1994). *Paralelização
do método de Gauss-Jordan de resolução de sistemas
de equações lineares*. U.AL.

IMPLEMENTING TRAIN APPLICATIONS BASED ON THE DISTRIBUTED SHARED MEMORY PARADIGM

C. BAE [*,1] M. RAYNAL [**,2] G. THIA-KIME [***,2]

* KHRC (Korea High Speed Rail Construction Authority),
Kumhwa Bldg, 16th F, 949-1, Togok-Dong, Kangnam-gu, Seoul
135-270, Korea, CBAE@chollian.dacom.co.kr
** IRISA, Campus de Beaulieu, 35042 Rennes Cédex, France,
raynal@irisa.fr
*** IRISA, Campus de Beaulieu, 35042 Rennes Cédex, France,
thiakime@irisa.fr

Abstract:
A high speed trainset consists of a power car (or power cars) and trailer cars. Each car has its own processor which communicates via a network in the train. For dependability, most of data (objects) in memory are replicated on each processor. To design applications in the context of distributed systems, a simple and portable method consists in utilizing the Distributed Shared Memory (for short DSM) paradigm. On a group of Sun Workstations, an evaluation model has been implemented consisting in a set of hierarchical software layers to offer a DSM with two consistency criteria, namely sequential consistency and causal consistency. On top of this DSM, the application layer implements a typical train application software namely door management. In this train application software, focus was put on portability, dependability and safety issues.

Keywords: Consistency Criteria, Distributed Shared Memory, Fault-tolerance, Portability.

1. INTRODUCTION

In a high speed trainset, most of the data (objects) in memory are replicated on each processor for dependability. Processors communicate via a network. If one object is changed by one processor, the modification has to be transmitted to all other processors and processed in a consistent manner: for example, with sequential consistency, care has

to be taken to ensure that modifications are made in the same order on all processes.

To design such an application in the context of distributed systems, the Distributed Shared Memory (for short DSM) paradigm is convenient to use. The design is more simple. A site which desires to disseminate information, for example to modify an object, just performs a write operation (thus just considers the well known shared variables programming paradigm); the underlying DSM protocol ensures that information is propagated efficiently, safely and in a coherent manner. Applications are portable, a write or a read operation does not require, for the site processing it,

---

[1] This work has been made in cooperation with GEC ALSTHOM Transport (France) according to the contract between KHRC and GEC ALSTHOM Transport.

[2] This work has been made in the context of a research performed for common benefit of IRISA and GEC ALSTHOM Transport.

to know about the topology of the communication network, the number of sites, the time-out delay, etc. This is taken into account by the DSM. Moreover, at the system level, load balancing, progress migration and program port become transparent to the programmer thanks to the DSM.

On a group of Sun Workstations, an evaluation model has been implemented consisting of a set of hierarchical software layers to offer a DSM with two consistency criteria, namely sequential consistency [Mizuno et al, 94] and causal consistency [Ahamad et al, 95; Raynal et al, 95]. On top of this DSM, the application layer implements a typical train application software, namely the door management system. In this door management system, including train functions related to passenger safety issues, focus was put on portability, dependability and safety issues. An important focus has been put on portability by using well-known tools like PVM (Parallel Virtual Machine), C language and by the structure of the implementation. This implementation was divided into layers, from the Transport Layer (in ISO/OSI model) through group communication (atomic and causal broadcast), distributed shared memory (sequential and causal consistency) to train applications.

This work includes new research results: causal consistency in the domain of new DSM consistency criteria and group communication model [Hadzilacos et al, 93], and experience in implementing industrial software (train door control applications). Safety issues are also considered, based on new research results in fault-tolerant asynchronous systems, mainly developed on top of concepts designed by Chandra and Toueg [Chandra et al, 91] such as consensus and failure detectors. Consensus, a fundamental problem of fault-tolerant distributed computing, cannot be solved deterministically in an asynchronous system that is subject to even a single crash failure [Fischer et al, 85]. To circumvent this impossibility result, Chandra and Toueg have proposed a tool, namely unreliable failure detectors, and studied the problem of using them to solve consensus. We use these results to build a reliable broadcast to ensure safety of information dissemination in spite of process crashes. As our implementation is structured into layers, all upper layers benefit from this safety reinforcement.

2. THE MODEL

2.1 Introduction

Our model has been defined on the basis of a study of a current on-board computer system (OBCS) on the TGV high speed trainset. This OBCS uses many processor units interconnected via the trainset network. Main characteristics of OBCS are: each processor unit has its own memory where replicated data and local data are stored, and communications are processed in synchronous mode.

So, a distributed system, consisting of many processes interacting via a DSM was considered. A DSM is a shared memory built on top of a distributed system.

Basically, shared memory is implemented on local memories of processes: copies of shared data (object) are simultaneously present in local memories. Each process has a memory manager. Memory managers communicate to ensure consistency of copies of a same object. The level of synchronization between memory managers determines the level of consistency of the shared memory. Atomic consistency is the most known one; it was implemented by Li and Hudak [Li et Hudak, 89]. It is associated with the usual semantics linked to a centralized shared memory.

Are presented below the following consistency criteria: sequential and causal consistency, and described protocols implementing them using group communication, namely atomic and causal broadcast.

2.2 Group communication

For information dissemination in a distributed system, a convenient and efficient way is to use multicast primitives. Two kinds of multicast primitives were used, namely atomic and causal broadcast, depending on the degree of reliability desired and the level of synchronization required.

2.2.1. Atomic Broadcast

An atomic broadcast (ABCAST) primitive ensures a total order on message delivery: messages sent with this primitive are delivered in the same order to each processor (See [Hadzilacos et al, 93] for ABCAST implementation). Moreover, each message processing is atomic: a message is delivered to all processes or is delivered to no process.

More formally, atomic broadcast guarantees that:

(1) All correct processes deliver the same set of messages,
(2) All messages broadcasted by correct processes are delivered,
(3) No spurious message is ever delivered (a message is delivered only if it has been broadcasted by some process),
(4) Messages are delivered in the same order on all processes.

Any protocol implementing a decentralized AB-CAST needs for each broadcast at least three phases:

(1) the sender sends a message to every process indicating he intends to make a broadcast,
(2) every process answer to sender with some information (usually a local message counter) enabling the ordering of distinct messages,
(3) on basis of answers of all processes, the sender makes a deterministic decision (message ordering) and sends it to all processes.

This protocol is blocking in the sense that each sender has to wait for answers from all processes. Moreover, it is intrinsically reliable, answers from all processes act also as an acknowledgment.

2.2.2. *Causal Broadcast*

Potential causality [Lamport, 78] provides a natural ordering (causal precedence) on events in a distributed system where processes communicate by message passing. Let us consider two events a and b. By definition, a precedes b if:

(1) a and b happen on the same process, and a happens before b (program-order) or,
(2) a is sending of a message, and b is reception of this message or,
(3) There exists an event c such that: a precedes c and c precedes b (transitivity).

Causal ordering enforces, for every process, the ordering of receive events (by the precedence relationship) as well as their associated send events.

A protocol implementing CBCAST ensures a broadcast with causal ordering of messages. It is usually designed by utilizing time-stamping. Each message piggy-backs a time-stamp which will enable each receiver to causally order messages. A sending needs only one phase and is non-blocking (see [Ahamad et al, 95] for time-stamp based CBCAST implementation).

2.3 *Consistency criteria*

A shared memory system is composed of a finite set of sequential processes that interact via a finite set of shared objects. Each object can be accessed by read and write operations. A write into an object defines a new value for the object; a read allows to obtain the value of the object. For simplicity, we assume all values written into an object are distinct. Moreover, some parameters of an operation are omitted when they are not important. Each object has an initial value; it is assumed that this value has been assigned by an initial fictitious write operation.

A shared memory computation is a partial order of the set of read and write operations issued by the processes. A consistency criterion is expressed as a constraint that the partial order has to satisfy[3]. A protocol implementing a DSM with some consistency criterion has to ensure that all computations comply with the associated constraint.

Now, two consistency criteria are defined: sequential and causal consistency.

2.3.1. *Sequential Consistency*

Sequential consistency was proposed by Lamport in 1979 to define a correctness criterion for multiprocessor shared memory systems [Lamport, 79] A system is sequentially consistent with respect to a multiprocess program, if *"the result of any execution is the same as if (1) the operations of all processors were executed in some sequential order, and (2) the operations of each individual processor appear in this sequence in the order specified by its program"*.

This informal definition states that the execution of a program is sequentially consistent if it could have been produced by executing this program on a mono-processor system.

Various cache-based protocols implementing sequential consistency have been proposed in the context of parallel architectures [Mizuno et al, 94] In most of these protocols, every local memory contains a copy of the whole shared memory.

So, a basic protocol to implement sequential consistency consists in utilizing the ABCAST primitive:

- To process a write operation, a process simply ABCASTs a message including name of the object written and the new value,
- to process a read operation, a process simply returns the value of the local copy,
- at reception of a write message, a process updates the object with the new value.

Such a protocol, using ABCAST's total ordering property, guarantees a total ordering on all write operations, and consequently sequential consistency.

2.3.2. *Causal Consistency*

Causal consistency was first introduced by Ahamad et al [Ahamad et al, 95]. It defines a consistency criterion strictly weaker than sequential consistency, and allows a wait-free implementation of

---

[3] It is worth noticing that formal definitions of consistency criteria are based on very few (and simple) definitions, namely: partial order, linear extension, sub-order and legality.

read and write operations in a distributed environment (*i.e.*, allowing resource cheap read and write operations).

Causal consistency is based on the *potential causality* concept of Lamport [Lamport, 78]. With sequential consistency, all processes agree on the same linear extension (of execution history). In a causal history, two processes are not required to agree on the same order for write operations that are not ordered in this history. Otherly said, all processes agree on the same partial order. However, sequential views of different processes may be different.

Ahamad et al [Ahamad et al, 95] presents an implementation of causal memory and studies programming with such a memory. Actually, when considering read and write operations equivalent to receive and send events in message passing systems, causal consistency is equivalent, in the shared memory model, to causal ordering [Birman et al, 87] for the delivery of messages in the message passing model.

So, a basic protocol to implement sequential consistency consists in utilizing the CBCAST primitive

Such a protocol, using CBCAST's total ordering property, guarantees a causal ordering on all write operations, and consequently causal consistency.

3. APPLICATION

The shared memory paradigm was used to simplify train door management application design.

To ensure good portability of the implementation, it has been structured into four hierarchical layers:

- train application (client programs),
- distributed shared memory (sequential and causal consistency),
- group communication (multicast, atomic and causal broadcast),
- Transport layer (in an ISO/OSI model).

The global structure is shown in Figure 1.

The transport layer is given by utilizing the PVM environment which ensures the corresponding quality of service, namely reliable communication. Moreover, PVM ensures some amount of failure detection of process crashes and communication link cuts.

Group communication services include atomic broadcast and causal broadcast.

Replication services include shared memory protocols implementing sequential and causal consistency. Sequential consistency which ensures a total order on update operations in a reliable

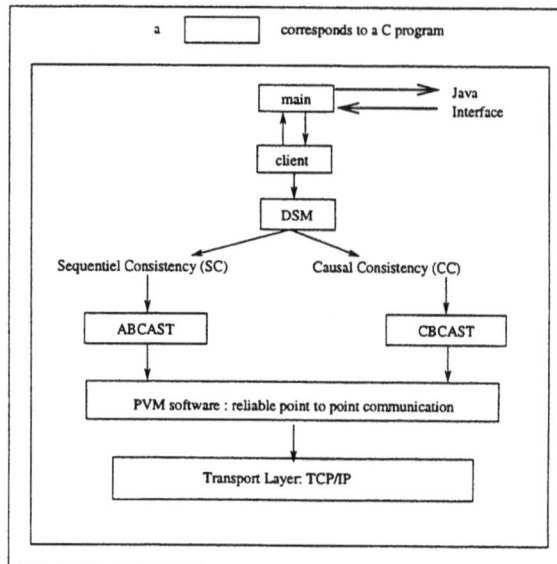

Fig. 1. Application structure

manner suits application software which uses critical data while causal consistency which can be implemented in a more efficient way suits application software which uses non critical data. These protocols are composed of 4 main procedures (scwrite, scread, ccwrite, ccread) corresponding to sequential write and read, causal write and read operations. Read operations check communication channels on which write messages may have arrived and deliver the value of the local copy. Write operations make a broadcast (atomic for scwrite, causal for ccwrite) of a message including name of object written and new value and update local copy.

Typical door management train applications consist of: signaling, operation mode, opening, closing, blocking control, obstacle detection, speed measurement, emergency handling, lightening and fault management. There is one PVM management application (main) and 20 train applications (clients) run by PVM processes executing in parallel.

4. MEASUREMENTS

Some measurements were made on a set of Sun Sparc 5 workstations running under Solaris linked by a 10 Mb/s ethernet network. Measurements of execution time were made from the beginning of the execution to include latency and bandwidth aspects. Thus, in all measurements, execution time includes an initialization phase.

The 20 processes are distributed by PVM on the set of workstations (the number of workstations is chosen statically before the beginning of the execution)

Two parameters were considered, first one is the number of workstations; second one is the number of writes processed by the application.

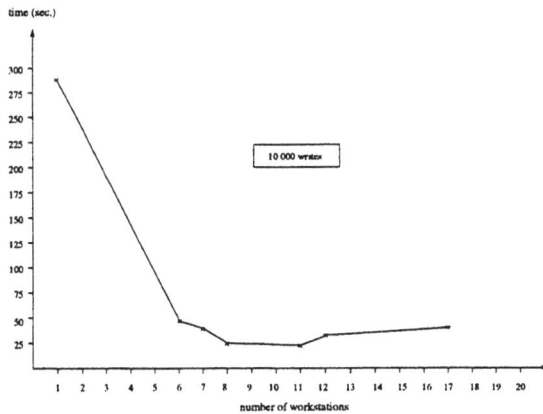

Fig. 2. Varying number of workstations

First, the number of workstations was varied (see figure 2), while considering 10 000 writes, and measured the time of execution. In figure 2, a good scalability of the algorithm is shown in regard to the number of workstations (for numbers up to 11 workstations). From 1 to 11 workstations, execution time decreases: process management and inter-workstation communication increase less than cpu computation delay decreases. This 11 workstations value is to be related to the number of processes (20). With more than 11 (*i.e.*, 1 process per workstation), execution time increases: time lost in process management and inter-workstation communication become greater than cpu delay.

Second, the number of writes was varied (see figure 3), while considering 5 workstations, and time of execution was measured.

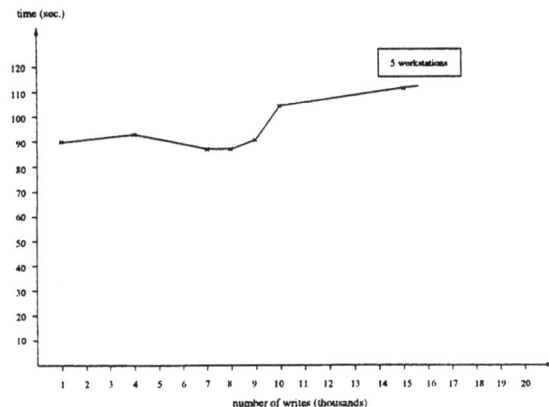

Fig. 3. Varying number of writes

In figure 3, a good scalability of the algorithm was shown in regard to the number of writes. First, note that there is some latency, a few writes need a non-zero time delay, due to the initialization phase. Then, execution time increases in a linear way according to the number of writes.

5. CONCLUSION

In this paper, the Distributed Shared Memory paradigm and consistency criteria, namely sequential and causal consistency, were described. In addition, an implementation consisting in a train door management application based on DSM was described. Reliability and portability issues have been also considered.

This implementation has brought to light a number of results:

It was shown that in such a hierarchically structured system, reliability at upper layers level of the system can be improved simply by enforcing reliability in lower layers (replication or group communication services layers).

The actual implementation of distributed door management was very much simplified by using Distributed Shared Memory paradigm. Actually, programming was almost similar to sequential (centralized) programming except that different processes are simultaneously executed and are sharing objects. Distributed programming was also simplified by the choice of using PVM to manage distributed processes. PVM offers an environment for distributed programming, namely there is a main program which spawns parallel client programs (20 client programs in our implementation).

To conclude, current work is focused on reinforcing reliability by studying group communication layer in more detail. The graphic interface issues is also studied by using Java Applets which ensure also some portability (as they are loadable by web browsers).

Acknowledgments: The authors wish to thank Gérard Demars (Gec Alsthom Transport) for his help during this work.

6. REFERENCES

[Ahamad et al, 95] M. Ahamad, P.W. Hutto, G. Neiger, J.E. Burns, and P. Kohli. – Causal Memory: Definitions, Implementations and Programming. *Distributed Computing*, 9:37-49, 1995.

[Birman et al, 87] K. Birman et T. Joseph. – Reliable communications in the presence of failures. In *ACM Trans. on Computer Systems*, vol. 5, number1, 1987, pp. 47-76.

[Chandra et al, 91] T.D. Chandra and S. Toueg. – Unreliable Failure Detectors for Asynchronous Systems. *Proc. 10th ACM Symposium on Principles of Distributed Computing*, Montreal, pp. 325-340, 1991.

[Fischer et al, 85] M.J. Fischer, N.A. Lynch, and M.S. Paterson. – Impossibility of Distributed Consensus with one faulty process. *Journal of the ACM*, 32(2):374-382, April 1985.

[Hadzilacos et al, 93] V. Hadzilacos and S. Toueg. – Fault-tolerant Broadcasts and Related Problems.

Chapt. 5, Distributed Systems (2nd Ed.), S. Mullender Ed., Addison-Wesley and ACM Press, 1993, pp. 97–145.

[Lamport, 78] L. Lamport. – Time, Clocks and the Ordering of Events in a Distributed System. *Communications of the ACM*, 21(7):558–565, 1978.

[Lamport, 79] L. Lamport. – How to Make a Multiprocessor Computer that Correctly Executes Multiprocess Programs. *IEEE Trans. on Computers*, C28(9):690–691, 1979.

[Li et Hudak, 89] K. Li and P. Hudak. – Memory coherence in shared virtual memory systems. In *ACM Trans. on Computer Systems*, vol. 7, number4, 1989, pp. 321–359.

[Mizuno et al, 94] M. Mizuno, M. Raynal, and J.Z. Zhou. – Sequential Consistency in Distributed Systems. *Proc. Int. Workshop "Theory and Practice in Dist. Systems"*, Dagstuhl, Germany, Springer-Verlag LNCS 938, (K. Birman, F. Mattern and A. Schiper Eds), pp. 227-241, 1994.

[Raynal et al, 95] M. Raynal and A. Schiper. – From Causal Consistency to Sequential Consistency in Shared Memory Systems. *Proc. 15th Int. Conf. FST&TCS (Foundations of Soft. Tech. and Theor. Comp. Science)*, Bangalore, India, Springer-Verlag LNCS 1026, (P.S. Thiagarajan Ed.), pp. 180-194, dec. 1995.

A DISTRIBUTED CONTROL SYSTEM ARCHITECTURE FOR STRIP-CASTER CONTROL & STORAGE YARD AUTOMATION

Duk-Man Lee, * Jin S. Lee, ** Tae-Wook Kang, *** and Joo-Kang Lee ****

* Instrumentation & Control Research Team, Pohang Iron & Steel Co., Ltd.,
Pohang, 790-785, Korea.
e-mail:pc554173@smail.posco.co.kr
** Department of Electrical and Electronic Engineering, Pohang University of
Science and Technology, Pohang, 790-784, Korea.
*** Research Institute of Industrial Science and Technology, Pohang, 790-784,
Korea.
**** Instrumentation & Control Research Team, Pohang Iron & Steel Co., Ltd.,
Pohang, 790-785, Korea.

Abstract. This paper presents an open architecture distributed control system using VMEbus based single board processors and PLCs (Programmable Logic Controller) as load-sharing partner. Taking advantage of their own features, we can design more robust and flexible system than those built with VME processors or with PLCs only. Based on this design architecture, the design guidelines about system bus, networking, MMI (Man-Machine Interface), DDC I/O, etc. are proposed to support high speed control and network-throughput required for real applications of POSCO such as strip-casting and thick-plate storage yard automation. These systems are now implemented and show good performance.

Keywords. Open-architecture, VMEbus, PLC, DCS, MMI, Strip-caster, Thick-plate storage yard.

1. INTRODUCTION

The environment for steel-making industries is harsh and full of dusts, noises, high-power signals, humid, and hot. Even under these harsh conditions, the controller must be robust for reliable operation. In addition to this, a number of sequence control functions are required in control problems to prepare for operation condition, to protect unusual operation, or to implement simple control routines, etc. In these respects, PLCs are the best candidate because they are reliable and are proven by many industrial applications. On the other hand, single board computers (or general purpose control board) are superior to PLCs because it is able to accommodate complex control algorithms and easy networking facilities, supports high level languages, depends much less on the vendors, etc. Thus, we present in this paper an open system architecture that incorporates single board computers and PLCs as load sharing partner. The basic idea of this architecture is to combine the advantages of each processor in the control environment.

In the main sections, two real application systems in POSCO (Pohang Iron and Steel Cooperation) are introduced: one is the new steel-making process called strip-casting and the other is the crane automation system for the thick-plate storage yard. The two systems use the proposed architecture as basic control configuration.

In designing the system architecture, the specifications for the overall system components such as system bus, DDC networking, MMI, DDC I/O, etc. are considered and the implemented results are explained in detail.

This paper consists of four sections. Section 2 describes design specifications used for the proposed applications. Section 3 introduces two application examples and explains each components of the systems implemented. Finally, section 4 concludes the paper.

2. SYSTEM DESIGN GUIDELINES

From oil industry to steel industry, from paper making industry to chemical processing industry, major portion of the process control systems that are currently under operation in Korea have been commissioned in the late seventies. As demands of the customer become ever increasing and more demanding, many of these systems apparently require partial updates or even complete renovations. In upgrading these systems, process engineers are looking beyond the traditional turnkey distributed control system (DCS) to open systems that incorporate open backplane, I/O, network, and software architectures. In doing so, they must be ready to accept a greater portion of the system integration hassles, but they obtain a number of benefits, particularly the high performance, increased flexibility, lower cost, and reduced single-vendor exposure (Marrin, 1993).

Especially, POSCO introduced a lot of turnkey based control systems from foreign makers. Since those are usually come from different makers and there is no unified specifications about system, upgrading system software and hardware and interfacing with other maker's systems are very difficult. Thus, in order to reflect the standard and efficient system architecture on the designing period and to have makers follow to unified rules, POSCO recently set up an EIC (Electric, Instrumentation and Computer) unification specification (POSTECH, 1995).

Referencing this guidelines and considering control characteristics of proposed applications which need high control frequency, the following points are carefully considered in designing the system architecture: the system bus should be an open architecture; the high-power and low-power I/O should be handled separately for the safety of system; the MMI should be easy to use and be GUI (Graphic User Interface) based; the networking method between DDC (Direct Digital Control) and MMI (Man Machine Interface) should be open and fast if possible; the networking between bus to bus does not affects bus bandwidth; the control boards installed in the bus must maintain its bus-access frequency as minimum as possible to share bus effectively.

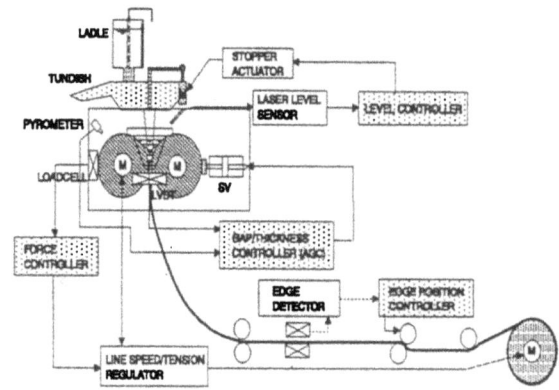

Fig. 1. Schematic layout of strip-caster pilot plant constructed in POSCO

3. APPLICATION EXAMPLES

3.1 Strip-Casting

Recently, the research and development of strip-casting system technology has gained much interest as a new type of casting method for the 21st century's steel-making process. The idea of strip-casting has emerged from the constant desire to produce hot-rolled thin steel strip directly from the molten steel, thereby simplifying the steel-making process. The method renders unnecessary the separate reheating and hot rolling processes, which require tremendous energy and operating costs (Lee et al., 1996).

Fig.1 shows the pilot strip-caster plant that has been constructed by Pohang Iron & Steel Co. (POSCO) and DAVY International Co. based on a twin-roll system which is similar to Bessemer's (Lee et al., 1996; Edwards and Willis, 1992). The pilot plant is about 60 m in length, and is designed to produce 2 - 6 mm thick steel plates.

Naturally, this complex system is equipped with many control units such as the mill drive control unit, the cooling control unit, the discharge control unit, the coiler control unit, etc. The mill drive control unit consists of several sub-control units: the molten steel level control unit between twin-roll cylinder, AGC (Automatic Gap Control) control unit, the edge-dam control unit, the loop-height control, and the rolling force control unit.

The overall control systems for strip-casting machine are shown in Fig 2. In this system, the standard VMEbus is adopted as system bus (Peterson, 1992). Since there are thousands of products, manufactures, software and bus interface chips available in the market, the vendor dependency becomes low, overall cost is down, and system upgrade will be easily done. This is the main reasons why the standard VMEbus is adopted.

Based on VMEbus, several single board computers are used for the mill drive control units and

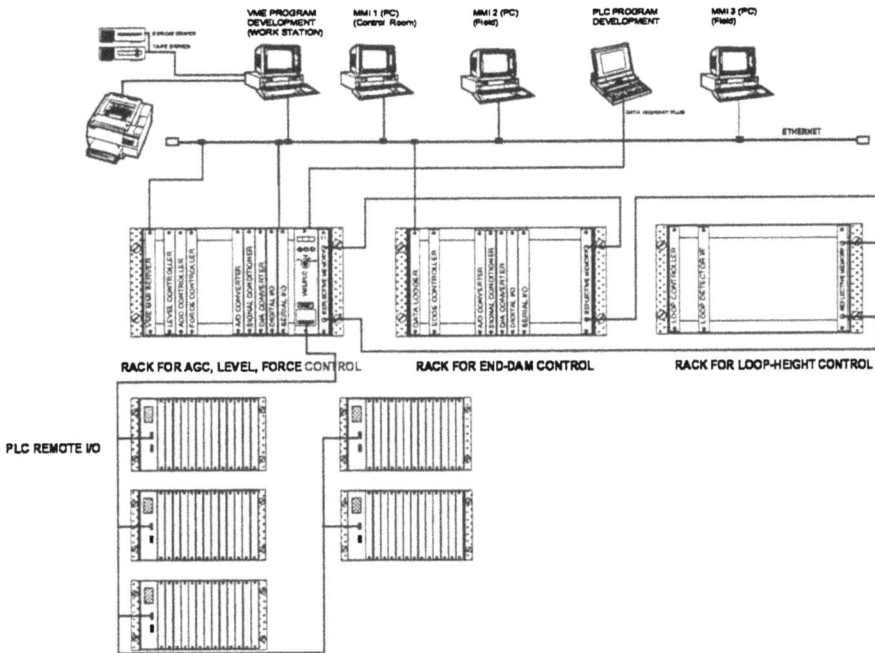

Fig. 2. Controller layout for the strip-caster

PLC is used for the cooling control unit, the discharge control unit, and the coiler control unit, where single board computers execute high-level control algorithms whereas PLC executes sequence control, high-power I/O manipulation, safety action, etc. To make the communication between PLC and single board computers easy, a rack-mounted PLC (VME-PLC) is adopted. Thus, the shared data can be transferred easily via high-speed backplane bus.

The application of a process control computer requires a series of interfaces for interacting between the computer system and the experts preparing, testing, integrating, operating, and maintaining its software and hardware. Such interface have been called man-machine interface. In the strip-casting application, the MMI is designed to run in the separate IBM compatible personal computer (PC) with windows operating system. The reasons why PC's are adopted as MMI stations are that the tasks in MMI station are not time-critical, the operator is familiar with PC, and the PC has a variety of application software and library. There is no control algorithm implemented in PC but the following interfaces are implemented: computer-operator interface for generation, test, documentation, and maintenance of system software; plant operator interface for monitoring and operating the plant at different hierarchical levels; production monitoring interface for plant management personnel (Popovic and Bhatkar, 1990). As a MMI design tool, a commercial software on windows environment are adopted and for high performance communication between DDC and MMI, ethernet and TCP/IP protocol is used.

Next, in order to confirm fast and reliable data transfer in multiple VME rack, the bus networking boards using the reflective memory concepts are employed. These boards which have local static RAM are installed in each VMEbus backplane, and all boards are connected by fiber-optic or flat cable with token-ring method. If a VME master board writes some data in the particular memory area of the networking board, the data transfer automatically occurs among the boards linked in the network. Actually, this networking does not affects bus bandwidth, and the memory contents of all boards in the network are same. Thus, this kind of networking board is highly necessary when the control racks are distributed and the control tasks running in each rack need to share information tightly.

In AGC and edge control routines, there are a lot of fast and time-critical tasks. Thus, if VME I/O boards are used for that tasks, there are much bus traffic in the backplane thereby it is difficult to share bus effectively among boards. In order to minimize bus-access and to share bus effectively in a high traffic bus, the single board computer which supports piggyback type I/O cards is selected. Since piggyback I/O cards only use local bus of single board computers, it does not affect VMEbus bandwidth. Recently, many vendors are announcing several piggyback type products and are easily available in the market.

3.2 Crane Automation System for thick-plate storage yard

The exact sensing and control of thick-plate number to lift and the fine control of crane to proper

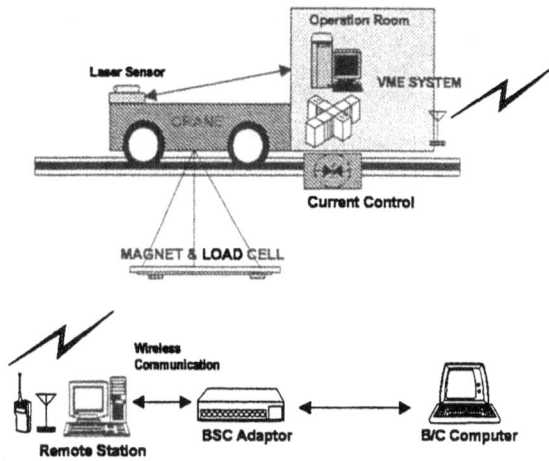

Fig. 3. Layout of crane automation system

position are essential for the thick-plate storage automation of POSCO. Previously, these actions are handled by two operators: one is overhead crane operator taking in charge of lifting the plates; the other is plate-counting operator standing in the ground and signals to crane operator how many plate are lifted. This kind of operation contains a lot of problems: without signaling man, the crane operator is difficult to know how many plates are actually lifted when he applies currents heuristically to the lifting electromagnet; there is no systematic method applying current profile to the magnet to lift the desired number of thick-plates; the signaling man is surrounded by much dangerous and noisy environments; the operation cost is very expensive, etc. These have been remained as unsolved problems in POSCO for many years. Thus, from 1992, POSCO and POSTECH have been developing an automation system for thick-plate storage house to realize one-man control system, where a load-cell and a magnetic flux sensor are introduced to accurately measure the desired number of plate, and an intelligent control algorithm to generate current profile for each plate combinations are developed. This project has been successfully completed and POSCO is currently going to invest 40 million dollars to accomplish one-man control operation in No. 2 and No. 3 thick-plate storage houses (POSTECH, 1996).

In this part, since rather detailed explanations about system components are given in the case of strip casting system, only overall system architecture and software configuration of this system are described simply. The control system layout is shown in Fig. 3. With similar architecture to strip-casting system, VMEbus and single board computers and PLC are used in this application. The controllers are located in the crane-operating room and the data transfer with remote supervisory computer is done via wireless modem. As Fig. 4 shows, five main functions are distributed in the control unit. Current control of magnet, high-

power signal interface, and emergency actions are given to PLC. The position control of crane with laser sensor data and the high-level algorithm to lift the desired number of plate are assigned to VME single board computers. MMI is implement in PC and the LCD is used for display due to severe vibration and steel dust exist in the crane-room. The communication between MMI station and VME system is done via RS422 line.

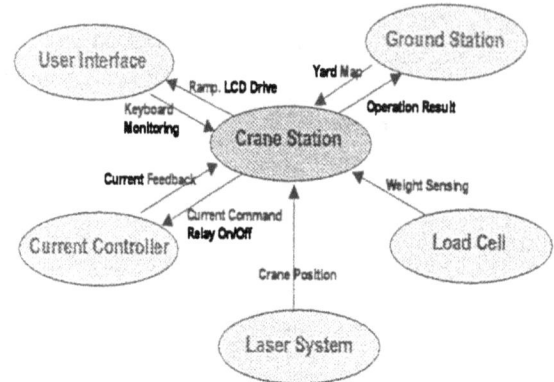

Fig. 4. Task configuration for crane automation system

4. CONCLUSION

In this paper, an open system architecture of distributed control system is proposed using single board computers and PLCs as load-sharing partner. Also, to guarantee control performance in a high-speed control loop, several techniques regarding system bus, networking, MMI, DDC I/O, etc. are explained with the applications of strip-casting and thick-plate storage yard automation.

The DCS fields are now extending its area to various fields and a number of good products are available in the market. However, it is true that no manufactures have all the resources to build the drivers to other systems that the customers demand. Therefore, the openness of any system is the doorway for the future success and protection of their technological advantages in their respective fields. The system designers and decision makers should consider this point seriously when they design the system architecture.

Acknowledgements

The authors wish to acknowledge the generous support from POSCO, RIST, and KOSEF through the ARC in POSTECH.

172

5. REFERENCES

Edwards, J.B. and I.C. Willis (1992). *A Simulator For A Continuous Strip Casting Machine*. Final Reports to Davy Distington on R & D Project.

Lee, Duk Man, Jin S. Lee and Taewook Kang (1996). Adaptive fuzzy control of molten steel level in strip-casting process. *Control Engineering Practice IFAC* 4, 1511–1520.

Marrin, Ken (1993). Inland steel turns to vmebus for distributed control. *Computer Design*.

Peterson, Wade D. (1992). *The VMEbus Handbook Third Edition*. A VITA Publication.

Popovic, Dobrivoje and Vijay P. Bhatkar (1990). *Distributed Computer Control for Industrial Automation*. Prentice-Hall.

POSTECH, E.E. Dept. (1995). *Survey Study to EIC Integrated System*. Technical Reports to POSCO on R & D Project.

POSTECH, E.E. Dept. (1996). *Automation software development of crane-system for thick-plate storage yard*. Technical Reports to POSCO on R & D Project.

AUTHOR INDEX

www.ingramcontent.com/pod-product-compliance
Lightning Source LLC
Chambersburg PA
CBHW081149250326
R18032300001B/R180323PG41598CBX00002B/3